The North America Tapestry of Time and Terrain

Cobertura de Tiempo y Terrenos de Norte América

L'Amérique du Nord : un collage de terrains d'âges différents

Tapestry Compiled by

Kate E. Barton[1], David G. Howell[1], and José F. Vigil[1]

Geology Compiled by

John C. Reed, Jr.[1] and John O. Wheeler[2]

[1]U.S. Geological Survey, [2]Geological Survey of Canada

2003

Sedimentary Rocks
Rocas sedimentarias
Roches sédimentaires

Volcanic Rocks
Rocas volcánicas
Roches volcaniques

Plutonic Rocks
Rocas plutónicas
Roches plutoniques

Metamorphic Rocks
Rocas metamórficas
Roches métamorphiques

Produced by U.S. Geological Survey, Map Distribution Center, Box 25286, Federal Center, Denver, Colorado, 80225, or 1-800-ASK-USGS. Available on the Web at http://geopubs.wr.usgs.gov/i-map/i2781. Manuscript approved for publication on December 4, 2002.

North America *through time:*

A PALEONTOLOGICAL HISTORY OF OUR CONTINENT

Lynne M. Clos

Fossil News
Boulder, Colorado

Published by
Fossil News
1185 Claremont Dr.
Boulder, Colorado 80305-6601
www.fossilnews.com

Text, photography, and illustrations by Lynne M. Clos
Book design by Lynne M. Clos

This book was produced entirely on Macintosh computers — an iBook and a G5 Desktop. The text was set in 12 pt. Beagle, the captions in 12 pt. Bernie Bold, and the chapter headings in 24 pt. Ameretto. Adobe InDesign software was used for the layout.

ISBN-10: 0-9724416-4-6
ISBN-13: 978-0-9724416-4-3

Library of Congress Control Number 2007936286

Printed on acid-free paper

Printed in South Korea

Four billion, six hundred million years ago...

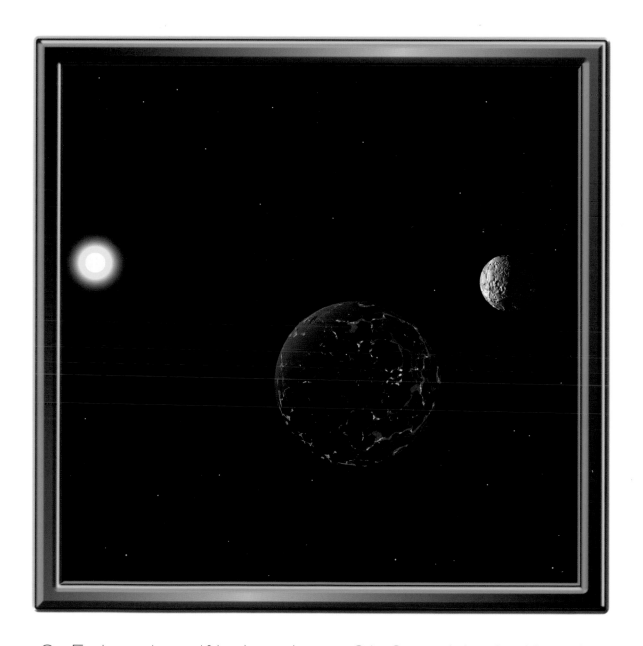

Our Earth was a churning, lifeless planet in the nascent Solar System, which condensed from a cloud of interstellar gas and dust. Yet it contained all the elements — chemistry, distance, time — necessary for life to evolve.

Geologic Time Scale

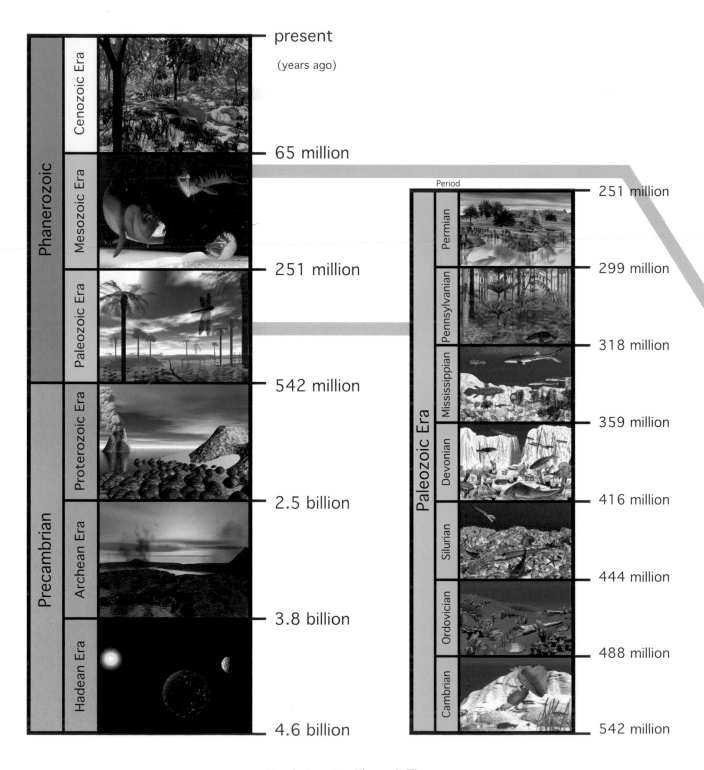

present

(years ago)

65 million

251 million

542 million

2.5 billion

3.8 billion

4.6 billion

Phanerozoic

Cenozoic Era

Mesozoic Era

Paleozoic Era

Precambrian

Proterozoic Era

Archean Era

Hadean Era

Period

251 million

299 million

318 million

359 million

416 million

444 million

488 million

542 million

Paleozoic Era

Permian

Pennsylvanian

Mississippian

Devonian

Silurian

Ordovician

Cambrian

North America Through Time

Dates from International Commission on Stratigraphy,
www.stratigraphy.org/gssp.htm, 2004

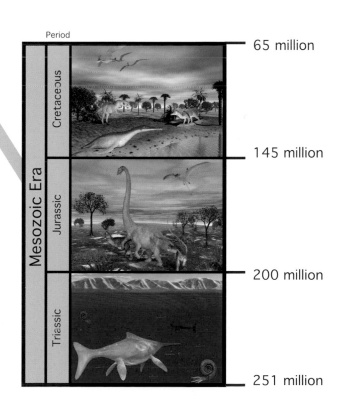

Period

Mesozoic Era

Cretaceous — 65 million

Jurassic — 145 million

Triassic — 200 million

— 251 million

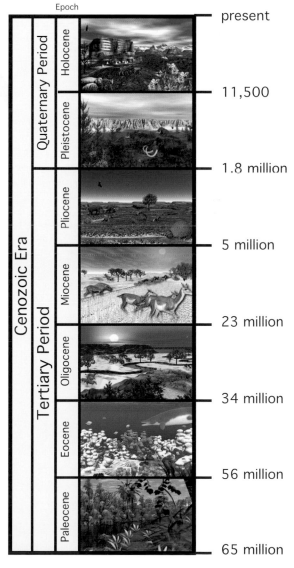

Epoch

Cenozoic Era

Quaternary Period — Holocene — present

Pleistocene — 11,500

— 1.8 million

Tertiary Period — Pliocene — 5 million

Miocene — 23 million

Oligocene — 34 million

Eocene — 56 million

Paleocene — 65 million

North America Through Time

Contents

Precambrian Time

Volcanoes bubbled and exploded, the atmosphere was poisonous, the oceans nearly froze over. Life evolved in a world very different from our own, and in so doing, transformed our planet. For nearly 90% of its history, life consisted of simple, single-celled organisms like bacteria and algae. We find them as fossils (really!), and they are responsible for making Earth inhabitable for the first animals. Evidence is found in peculiar rocks called 'banded iron formations' 5

The Cambrian

From small, strange, soft-bodied creatures that lived on the seafloor, in a span of ten million years almost all the modern phyla with easily-preservable hard parts had evolved. Probably the others did too, but have not left a fossil record. This is called the 'Cambrian Explosion.' What can account for this sudden burst of evolution...or is it really an artifact of preservation? And why have no new archetypes arisen since then? 21

The Ordovician

The greatest radiation of animal families ever to take place occurred during the Ordovician Period. Although the basic body plans were already set, Nature took this opportunity to produce endless variations on these themes. Life was still restricted almost entirely to the marine realm, but there are tantalizing hints that plants may have begun their colonization of the land. And the Earth experienced a major episode of glaciation, wreaking havoc with the environment...37

The Silurian

The greening of the land began apace with the establishment on land of simple, moisture-loving plants like mosses and liverworts. In the marine realm, trilobites reigned supreme and were hunted by voracious, predatory nautiloid cephalopods and eurypterid 'sea scorpions.' North America occupied tropical latitudes, and the climate returned to a "greenhouse" one

The Devonian

Fishes diversified into an enormous array of shapes and sizes, and the largest of them became fearsome predators. The first large trees appeared on land, and the earliest forests arrived on the scene, tempting some animals to leave the water in search of new habitats. A major extinction event occurred late in the Period, decimating marine invertebrates

The Mississippian

The world was warm, sea levels were high, and large areas of North America were covered by shallow waters. Coral-sponge reefs formed on the continental shelves and have left us huge deposits of marine limestone. Forests of crinoids — animals that looked like underwater flowers and have been nicknamed 'sea lilies' — grew everywhere, and in some places their fossils make up most of the rock

The Pennsylvanian

The waters receded from much of our continent, and vast swampy lowlands covered most of the land. Forests of fern, horsetail, and clubmoss trees grew everywhere; their remains form the coal deposits of the Appalachian region, and much of the world. Huge carnivorous amphibians lurked in vegetation-choked rivers and made forays onto land, perhaps to sun themselves after a good meal

The Permian

Truly dry-land-adapted animals became commonplace: the reptiles. No longer were they tied to water for their reproduction; their young could develop fully in a terrestrial environment, safe inside a shelled egg. Plants, too, invaded drier habitats, relying on seeds for reproduction instead of spores. Climates the world over were becoming more arid as all the continents assembled into a single supercontinent — Pangea, which brought with it extensive deserts and strong monsoons. The Period, and the Paleozoic Era, ended with the greatest mass extinction of all time, which may have been caused by extensive volcanism and the sudden release of vast seafloor methane stores

The Eocene

The "dawn of the recent" saw the last great time of tropicality for North America. Rainforests cloaked much of our continent and palm trees grew in Greenland. Most of the modern orders of mammals appeared at this time, including another important native North American group, the horses. Bizarre uintatheres with knobby horns on their faces browsed on the lush vegetation, and flight evolved for the fourth and final time

The Oligocene

Antarctica settled firmly over the South Pole, severing its ties with the other southern continents and redirecting ocean currents. This ushered in a slow decline in temperatures that saw the great tropical forests retreat south of the border. Animals, and plants, had to adapt to cooler and drier conditions; as the vegetation became more open, running became a useful skill

The Miocene

Grasslands spread across the interior of our continent as the rains became less reliable. Herds of horses and camels made use of this new forage. As they came to rely more and more on the abrasive grasses for food, their teeth began to change as well, becoming higher-crowned to survive a lifetime of wear. Rhinos browsed in the shadow of the Cascade volcanoes, which belched huge clouds of ash high into the North American skies, creating some amazing fossil bonebeds

The Pliocene

Lowered sea level, from the buildup of Antarctic ice, connected us via Alaska to Asia, allowing hairy elephants and hordes of other invaders to enter North America. The Panamanian Land Bridge joined us to our southern neighbors, allowing giant armadillos to come north and other animals to go south. It continued to get colder

The Quaternary Period

The Pleistocene

Finally, the ice came. It crawled down from Greenland and Hudson Bay to cover our continent as far south as Ohio. Woolly mammoths grazed on lush grasses and forbs at the edge of the ice; mastodons browsed in spruce swamps across the Great Lakes region; sabertooth cats and packs of dire wolves hunted bison, pronghorn, and deer. Time and again the ice advanced and retreated. When last it melted, most of the large mammals vanished with it. Why? Was it related to the appearance of a new mammal on our continent — *Homo sapiens?*

The Holocene

North America's large mammal fauna is severely depauperate compared with what it has been in the past. Will the trend of extinctions — at our hand — continue? And the Ice Age is not over. We are merely living in an interglacial spell between inevitable advances of the ice. But paradoxically, our world is warming, and again we are responsible. Can we — and should we — try to stop it? What lessons can we learn from North America's past that might help us decide? 243

 *this book is dedicated to*

Athena,

 *Osiris,*

 *and Moe,*

who kept me company

through the long hours and late nights

of its creation,

and to the memory

of Rooter

 *and Clide*

About the Author

Lynne M. Clos holds a Master's degree from the University of Colorado in museum studies with a specialization in vertebrate paleontology. Her Master's research focused on a new Miocene varanid lizard from Kenya and relationships within the Varanidae. A member of the Society of Vertebrate Paleontology, the Association of Applied Paleontological Sciences, and the Western Interior Paleontological Society, she has done fieldwork in such far-flung places as the North Slope of Alaska, Arizona's Petrified Forest National Park, Porcupine Cave high in Colorado's Rocky Mountains, and cliffs along the seashore of South Australia. She formerly worked at the Denver Museum of Natural History explaining the dinosaurs of the Ghost Ranch quarry to the public. Since 1998 she has been editor and publisher of *Fossil News* (www.fossilnews.com), a monthly journal for avocational paleontologists. Her paleoart is on display at the *T. rex* Museum in Tucson, Arizona. She received the Decree of Excellence in Literature in 2005 from the International Biographical Centre in Cambridge, England, and was featured in Two Thousand Notable Women in 2003. Her first book, Field Adventures in Paleontology, was published in 2003. She lives in Boulder, Colorado with her husband Chris, daughters Allison and Mattie, dog Moe, and numerous fish, including two silver arowanas named Osiris and Athena.

Preface

Sometime during the spring of 2003, when I was finishing up my first book, *Field Adventures in Paleontology*, my friend Allen made a suggestion for a series of articles he thought I might like to run in the journal I publish, *Fossil News*. His idea was to do a piece on each of the geological time periods, and what was going on paleontologically during that time. I liked the idea, but thought it would benefit from a narrower focus. I thus decided to write the articles with what was happening in North America playing a central role.

I ran the series the following year, beginning with the Precambrian and ending with the Quaternary. I illustrated the pieces with photos of North American fossils and with paleoenvironmental reconstructions I created on my computer using other photographs and a 3D modeling program that specializes in landscapes, called Bryce. It was so much fun that I decided that as time permitted, I'd like to turn it into a second book. I wanted to expand and rework the articles and artwork, and instead of treating the Tertiary and Quaternary periods as a whole, instead write a chapter on each of the individual Epochs comprising them.

Well, five years later, here it is. In introducing the book to you, I'd like to say a few words about who I've written it for. It's not intended as a textbook—although I'll be delighted if teachers find the material appropriate for their classes. I'm really writing for those of you who have an abiding interest in fossils and Earth's history, but who have not pursued a formal education in the field. High school classes in biology and geology (or equivalent reading) will be helpful; but I don't assume that you have an advanced degree, or even college coursework, in the earth sciences. There's plenty of professional literature for that. I do, however, assume that you're familiar with the basics—so that I don't need to explain to you what a trilobite is, or that continents drift about the globe. If that's what you're looking for, many other excellent books have been written for just that audience, and I suggest you pick up one or two. There is a glossary in the back of the present volume, but if you have to refer to it every sentence or two, it will limit your enjoyment of the book.

On the other hand, if you see yourself in the above description—welcome aboard! I hope to take you on an enjoyable and intellectually stimulating journey through the history of life on our planet and the role our favorite continent has played in it all. We'll cover how North America came to be, how life has evolved and our continent's role in it, and where in North America you can both view and collect some of these fossils. Although I have attempted to cover North America *sensu lato*—from the Arctic Ocean to the Panama Canal—the book concentrates on the U.S. and

Canada, for two purely practical reasons. One is that Mexico and Central America did not dock with the remainder of our continent until quite late in geological history, and have little exposed bedrock from times prior to the late Mesozoic Era. This limits what we know about the early development of lands south of the border. The second reason is that there has been far more geological fieldwork done north of the Rio Grande than south of it, so again, much more is known about the northern regions. This focus on the U.S. and Canada is in no way meant to imply that they are more important than the remainder of North America.

Finally, I'd like to thank several people for the roles they've played in helping me create this book. First is Allen Debus, who hatched the original idea, and has followed it as a magazine series through its metamorphosis into a book, offering many helpful suggestions along the way. Hans Hofmann was instrumental in pointing me towards invaluable published material on Precambrian fossils in North America. Many fossil collectors, most especially Marc Behrendt, have allowed me to photograph their collections and individual pieces. And my family—my parents, husband, and two daughters—have offered critique and encouragement, and also been infinitely patient as I dragged them out fossil collecting and to visit museums and parks from coast to coast in search of information and photographs. I couldn't have done it without you all.

Lynne M. Clos
Boulder, Colorado
summer 2007

The First Four Billion Years

The first rocks had solidified by 3.85 billion years ago, and the Earth had cooled sufficiently for water to condense into oceans. A sunset over the Archean sea reveals active volcanism, but in the oceans is the recipe for life.

Chapter One
Precambrian Time

Two billion years ago, the seashore is fringed with algal mounds called stromatolites. They were the most advanced lifeforms of their time, and are the most common macrofossils found in Precambrian rocks.

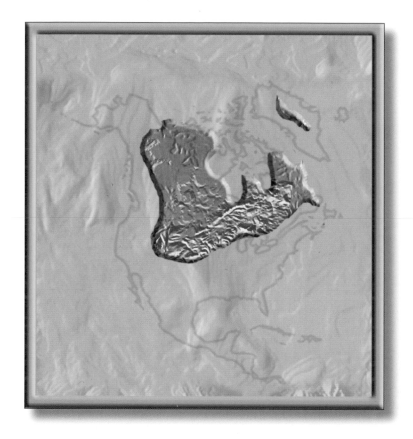

Precambrian North America

Our Earth is about 4.6 billion years old, approximately one-third as old as the universe. All the heavy elements in our solar system—carbon, silicon, iron, indeed all the elements but hydrogen— were formed in the furnaces of previous generations of stars, and flung into the far reaches of space when those stars exploded at the ends of their lives. Why is this important? Because without these heavier elements, our planet, and the life that dwells upon it, could not exist. We owe a great debt to the stars that have already come and gone.

Our solar system condensed from a nebula of interstellar dust and gas. Gravity pulled it together; clumps of dirt called planetesimals formed in the denser regions of the cloud; and the whole thing simultaneously shrank and spun, flattening itself out into a disc shape. Gradually the planetesimals accreted to form the planets of our solar system, orbiting the central Sun in mostly co-planar ellipses. The collision between the nascent Earth and a particularly large planetesimal (perhaps the size of Mars) probably flung a large blob of molten rock into orbit, creating our Moon. For about the first 800 million years of Earth's history (the Hadean Era, named for the hellish conditions prevalent then), our planet was nothing more than a molten ball of lava. It did not cool enough for the first rocks to form until somewhere around 3.8 billion years ago (bya).

In fact, we only know that by 3.8 bya, there were areas of solid rock in a few places. Some rocks may have congealed a bit earlier, but none of those survive to the present day. Recycled zircon inclusions in a few very old rocks hint at having crystallized around 4 bya, but the oldest rocks that have been found intact date to 3.85 bya. They are known from our continent—from Greenland, which is geologically a part of North America. Thus the history of North America is particularly significant for the history of our planet as a whole.

Definition and Nomenclature

The title of this chapter is "Precambrian time." When was the Precambrian? It isn't a time period in the strict sense of the word, like those we will cover in subsequent chapters; it's just "the Precambrian." Basically, it is everything previous to the start of the Cambrian (chapter 2) at 542 million years ago (mya). British geologist Adam Sedgwick defined the Cambrian in 1835, placing the base of it at the first appearance of the fossils that were known at the time. Hence the name Precambrian for everything that came before.

A lot has happened since 1835, and we now know that there was life (and fossil evidence of it) throughout most of the Precambrian. Geologists have divided the Precambrian into three Eras: the Hadean, the Archean, and the Proterozoic. The Hadean, as aforementioned, begins with the formation of our planet and ends 3.85 bya with the consolidation of the oldest known rocks. The Hadean need not concern us further here, because there was no life then, and this is a paleontological history. The Archean ("ancient time") begins with those 3.85 billion-year-old rocks from Greenland, and ends with the onset of the Proterozoic ("age of first life") some 2.5 bya. The event which defines the boundary between the Archean and the Proterozoic is the appearance of the oldest continental rocks which have not been reheated or chemically altered. The Proterozoic extends from 2.5 bya to the base of the Cambrian at 542 mya.

Quartz-magnetite iron formation, 3.8 billion years old, Isua, West Greenland. Although no fossils this old have been found, the presence of banded iron formations — the first evidence of photosynthesis — is suggestive of life. New Mexico Museum of Natural History, Albuquerque.

Typically, as I'll mention in each chapter in turn, "global stratotypes" (official geographic locations and specific rocks found there, defined by the International Commission on Stratigraphy) are designated for each division of the geologic time scale. There isn't one for the beginning of the Precambrian (due to lack of rocks!), but there is one for the end, at the Precambrian-Cambrian boundary. This is 2.4 m above the base of Member 2 of the Chapel Island Formation at Fortune Head, Newfoundland, Canada, at the first appearance in the rocks of the trace fossil *Trichophycus (Phycodes) pedum*, which approximately coincides with the appearance of the so-called "small shelly fauna." Divisions within the Precambrian (e.g. Archean, Proterozoic) are defined chronometrically, but without an "official" rock location. An exception is the Ediacaran Period, the last bit of the Proterozoic Era (and a time of particular importance to us, as we shall see). Whereas its beginning is only vaguely estimated as c. 600 - 650 mya (from bracketing radiometric dates), stratigraphically it is defined at the base of the Nuccaleena Formation cap carbonate, immediately above the Elatina diamictite in the Enorama Creek section, Flinders Ranges, South Australia. The end of the Ediacaran is the Precambrian-Cambrian boundary. Thus Precambrian time, as a whole, lasted for some 4 billion years, or almost 90% of Earth's history.

Paleogeography and Paleoclimate

Before launching into the evolution of life, let's examine how North America itself came to be. At the start, our planet was molten and smothered in noxious gases. The primordial atmosphere was composed chiefly of carbon dioxide, methane, ammonia, and nitrogen, and was without any free oxygen—totally unbreathable for a visitor from today. As the planet cooled, scum formed on the surface of the lava seas—scum composed of the lighter rock types, such as granite. But there were no continents, no oceans. Gradually, outgassing of rocks and the melting of ice from comets delivered enough water to form a global sea; however, there were no landmasses rising above the waters which compare with the continents of today. A vast mélange of terranes, islands of scum, perforated the ocean's surface, but they were relatively small and spread out every which where.

The North American continent (and indeed, the other continents as well) formed gradually as tectonic action splatted these small terranes up against each other and welded them together. The oldest portion of North America is made up of the rocks of the Canadian Shield, which form the stable core of the continent (termed a 'craton'). Geographically, the Canadian Shield is the central and northern region of North America—what you'd have if you removed the Rockies and everything west of them, the Appalachians, the Gulf Coastal Plain, and the entirety of Mexico and Central America (all of which would dock with our continent later, as we'll see.) Geologists have identified some of the separate terranes that comprise the Canadian Shield, which we may call by names such as Bearslavia, Churchill Land, Superior Land, South Superior, East Greenland, Labrador, and Grenvillia.

The fine details of the initial assembly of North America are difficult to decipher, and are best left to professional papers on geology, but some generalities may be outlined here. Around three billion years ago, Bearslavia bumped into Churchill Land, raising a range of high mountains. The weathered remnants of these mountains can be seen in Canada's Northwest Territories today. About the same time, elsewhere in the ocean, South Superior plastered itself against Superior Land, forming the South Superior Mountains in what is now the northern Great

Lakes region. Labrador docked from the east, creating another mountain range at the suture. (When I say 'south' or 'east,' that is in terms of their positions today; paleolongitude is not known.) These terranes would form the southern part of the Canadian Shield, stretching from Labrador to Manitoba. Churchill Land and Superior Land formed two separate mini-continents at this time. East Greenland was a separate island, as was Grenvillia to the south. This phase was complete by 2.5 billion years ago, or the end of the Archean.

As the Proterozoic dawned, between 2.5 and 1.7 billion years ago, Churchill Land and Superior Land were welded together, raising the lofty Churchill Mountains along the suture and assembling the central craton. Older mountains had been eroded down to hills. The Churchill Mountains occupied the region where Manitoba, Saskatchewan, and the Northwest Territories come together today. Then, a little over a billion years ago in what is called the Grenville Orogeny, Grenvillia splatted up against the south shore of Superior Land, pushing up the Grenville Mountains and completing the North American craton (sometimes also called Laurentia). The eroded core of the Grenville Mountains forms the hilly terrain of the St. Lawrence region of modern Québec.

The Grenville Orogeny is particularly significant, because it was part of the assembly not only of North America, but of a massive supercontinent called Rodinia. You may, from other readings, be familiar with the supercontinent of Pangea ("all lands") that existed much later (around 250 mya), and that we will mention in subsequent chapters. Well, that wasn't the first time that, in their ceaseless peregrinations, the continents had all assembled into one great landmass. In fact, it has been proposed that such things have happened perhaps half a dozen times in Earth's history, but evidence for very early supercontinents is scanty. Rodinia, which assembled around 1.1 - 1.0 bya, is the earliest supercontinent for which there is unassailable evidence. North America—at least, the Canadian Shield—formed the core of Rodinia, lying in mid- to high-southern latitudes and flanked by Australia and East Antarctica on the west, and Baltica and Amazonia on the east.

Two billion-year-old banded iron formation, Temagami, Ontario. Dark grey iron-rich bands alternate with red bands of iron-poor chert. Author's collection.

Supercontinents have great import for global climatic regimes. Rain clouds never travel very far inland across a continent without being wrung dry by passage over mountains: witness the arid interiors of modern Australia, Asia, and western North America. This effect is had in spades when a supercontinent exists.

Little rain, extensive deserts, little weathering of rocks: all are hallmarks of a super-continent, be it Rodinia, Pangea, or some other.

But supercontinents don't last; plate tectonics is restless—rifts form, tearing the continent apart from within; and the pieces, in various forms, drift their different ways. This, too, affects climate: once ocean waters gain access to what were formerly continental interiors, rainfall increases, and along with it, the weathering of rocks. Remember the carbon dioxide and methane in the atmosphere? Well, without them (and the "greenhouse" effect they engender, an all-too-familiar topic in the news today), global temperatures would be some 30°C (54°F) lower than otherwise—and the entire planet would be frozen over, oceans and all. Bad news for life. Amazingly, there is evidence that we came perilously close to that scenario some 700 mya as Rodinia broke up. The increased rock-weathering sucked CO_2 out of the air and locked it up in carbonates (mainly limestones), sending temperatures plummeting. Glaciers covered vast areas of the continents, much as they did some 18,000 years ago at the height of the last Ice Age. Some scientists have even suggested, based on computer models, that the entire oceans did freeze—a scenario termed "snowball Earth." Dramatic as this sounds, it is probably a little overboard. Tillites (a type of glacially-derived rock) occur in deep ocean sediments from mid paleolatitudes, indicating that icebergs were floating out over open water and dropping the sand and gravel they carried at their bases. This couldn't have happened if the entire ocean was frozen. Additionally, an effect called 'albedo' kicks in when large ice-caps form. Think of getting in your car on one of those bright-blue days right after a big snow has just passed. It's blinding, right?—and you probably have to wear sunglasses. That's because the albedo (reflectivity) of snow and ice is very high (due to their white color). Ice absorbs little sunlight: most of it is reflected back into space (or at your retinas). This effect sets up a positive-feedback loop whereby the more it snows—or, more precisely, the more of the Earth's surface area that becomes snow-covered—the less solar heat the planet retains, and the colder it gets. Such an albedo effect sustains an Ice Age once it begins, until astronomical factors increase the amount of incoming solar energy and break the cycle. This works fine with Ice Ages such as the current set, and partially accounts for why there is alternation of glacial and interglacial cycles. But if the entire surface of the Earth were to freeze over (and become white), in all likelihood (and according to other computer models), it would never be able to thaw again—the Sun's radiation just isn't strong enough (and during the time of Rodinia, is estimated to have been about 6% less than it is now, because stars become hotter as they age). Obviously, the Earth has not become permanently frozen over, suggesting that Rodinian Ice Ages were more similar to the current episode than to the "snowball" scenarios. Thank heavens!

Small polished stromatolite slab of unknown provenance. The layers can be clearly seen. Author's collection.

When Rodinia split, it cleaved into halves, creating the Iapetus Ocean. North America drifted across the South Pole, while Antarctica, Australia, India, South America, Arabia, and parts of China (which would, later, make up Gondwanaland) stretched northward almost to the opposite pole. The Congo craton—the core of Africa—drifted alone. In what is called the Pan-African Orogeny some 550 mya, the Congo got squished in the middle as the Gondwanan pieces of Rodinia reassembled into a new supercontinent, Pannotia. North America (Laurentia) remained solitary, and slowly drifted by its lonesome toward warmer latitudes, though still in the southern hemisphere.

Life in the Precambrian

Reports of genuine Precambrian fossils date back to the late nineteenth century, when explorers in the Hudson Bay region of Canada found strange laminated structures now known to be stromatolites (more about these in a minute). With the advent of modern microscopic techniques, the hunt for the oldest fossils has pushed the frontier far back in Precambrian time, into the early Archean. The oldest undisputed prokaryotic cells (those without a nucleus) are from rocks dating to 3.5 bya, but possible microbes, and chemical signatures that may be from living organisms, date back almost to the very beginning of the Archean Era at 3.8 bya. Clearly, once the planet had cooled to a reasonable temperature and acquired oceans, life wasted no time in getting started. Experiments show that most of the amino acids found in living cells can form quite easily from the inorganic compounds thought to have been present in the primitive oceans, but only in the absence of free oxygen.

Precambrian stromatolites, Belt Series, Glacier National Park, Montana. The mounded layers can be clearly seen. North American Museum of Ancient Life, Lehi, Utah.

Within short order, microbes appear in rocks from around the world. These first cells made their living as chemotrophs ("chemical eaters"), perhaps in a fashion similar to those found in deep-sea vent communities today. Remember, there was no oxygen in the atmosphere, so these were anaerobic microbes. It took the advent of photosynthesis to begin to generate free oxygen. The first photosynthetic organisms were called cyanobacteria or blue-green algae. By 3.5 bya, they had learned to live in colonies and form distinctive structures known as stromatolites. Archean stromatolites are known, but are uncommon; most stromatolites are from the Proterozoic. For most of the Precambrian, they are the most complex fossils we find.

A stromatolite is basically a cyanobacterial mat that forms regular, cyclic "bands" as it grows upward to stay in the sunlight. Twice each day, with the advancing tide, a layer of fine sediment blankets the bacterial mat. As the tide recedes, the cyanobacteria grow upward to retain their ability to photosynthesize. Eventually they form finely-laminated structures that are preserved in the rocks and are quite distinctive from abiotic laminates. When sectioned, stromatolites frequently present as lumpy, columnar structures. They can be quite beautiful when polished.

The heyday of stromatolites was the latest Archean and Proterozoic, from about 2.9 bya to 700 mya. These cyanobacteria, the inventors of photosynthesis, are responsible for changing the primordial atmosphere into the one we have today. In the absence of EPA regulations on toxic waste, they expelled the byproduct of photosynthesis—oxygen—into their environment. We think of oxygen as a necessity, but to these early lifeforms, it was a poison. In the process, they created some of the most distinctive Precambrian rocks on Earth: banded iron formations. Most of the world's iron ore is mined from banded iron formations, which generally contain 15% or more iron content. The largest of these deposits on Earth crops out in Michigan's Upper Peninsula, across Wisconsin into northern Minnesota, and up into Ontario.

Banded iron formations consist of thin (millimeter- to centimeter-scale) alternating layers of iron-rich magnetite, Fe_3O_4, or hematite, Fe_2O_3 (black or grey), and iron-poor chert or jasper (frequently reddish). Geologists have puzzled enormously over exactly how these banded iron formations were created—and why they appear almost exclusively in rocks 2.5 to 1.8 billion years old, but practically never younger. The most widely-accepted theory is this. Photosynthetic cyanobacteria clustered in the upper layers of the global oceans, releasing free oxygen into the water. Weathering of basaltic rocks had made the oceans rich in iron, which precipitates as iron oxides when it combines with oxygen. So, as the blue-green algae (cyanobacteria) flourished, sediment layers rich in iron compounds (the black hematitic layers) were laid down. In essence, the oceans began to rust! But the algae's success spelled their doom—eventually the oxygen built up above the level that the free iron could absorb, to such a concentration that it poisoned the algae—and their population crashed. Oxygen now became scarce. After this, a layer of chert (which is mostly silica, and contains only a little iron oxide) would slowly precipitate onto the seafloor. Gradually, the algal population recovered, and the cycle began again.

Two billion-year-old stromatolites, Biwabik Formation, northern Michigan. Variations in iron content provide the beautiful black and red coloration, which makes for a stunning slab when polished. Private collection.

This scenario accounts for both how the banded iron formations came to be, and for why they are no longer forming today. Eventually, nearly all the free iron in the oceans got used up. Today's oceans are almost totally devoid of iron, except in rare, anoxic (oxygen-poor or -free) places like the bottom of the Red Sea (where the seawater contains 5000x the iron that it does elsewhere). No more iron in the seawater—no more banded iron formations. Only after this had occurred could free oxygen begin to diffuse into the air. The stage was set for the evolution of aerobic life—organisms which could tolerate this "polluted" atmosphere. Because oxidative respiration releases as much as 18 times the energy of anaerobic fermentation, there was a big advantage for any cells which could learn to use the "polluted" air.

Medusinites asteroides, Ediacaran assemblage (cast). New Mexico Museum of Natural History, Albuquerque.

By around 1.8 bya, eukaryotic cells—those with nuclei—emerged on the scene. The prevailing theory as to how this came about is that larger bacteria engulfed smaller, specialized ones and put their talents to work in organized cells. Both the chloroplasts of plants and the mitochondria of animals are thought to derive from bacteria which were once free-living; those hardworking cyanobacteria, once sheltered inside eukaryotic cells, became the basis of photosynthesis in all of the plant kingdom. This was the first step on the road to multicellular life.

Somewhere around the same time, eukaryotes also invented sexual reproduction. Prokaryotes don't have it—they reproduce by simple division. The big advantage to sex is the reshuffling of genes, which provides variability in the population. A genetically homogeneous population (of anything) is extremely sensitive to annihilation from environmental perturbations; this is the worry with highly endangered species today (such as the few remaining cheetahs, which are all so much alike that they can accept skin grafts from each other). A genetically heterogeneous population is much more likely to possess some members who are resistant to any disease or environmental fluctuation that might come along, preventing the species from going extinct.

Dickinsonia costata, Ediacaran assemblage (cast). Denver Museum of Nature & Science, Denver, Colorado.

We noted earlier that stromatolites decline in abundance in the fossil record after about 700 mya. They didn't go extinct; a few of them survive today in hypersaline lagoons such as in the Exuma Cays in the Bahamas, and Shark's Bay in western Australia. (Obviously, these are species which are oxygen-tolerant.) But stromatolites are tasty, and no match for grazing animals. They only flourish in the aforementioned locations because the hypersalinity of the water excludes grazers like sea urchins and snails. Of course, complex, hard-shelled animals such as these weren't around yet (we'll meet those phyla in chapter 2), but *something* was eating the stromatolites, and was likely responsible for their decline.

Precambrian stromatolites, Mary Ellen Jasper, Iron Range, Minnesota. North American Museum of Ancient Life, Lehi, Utah.

Earlier in this chapter, we mentioned the Ediacaran Period—the last 60 - 100 million years of the Precambrian—and noted that it would be of particular importance to our story. This timeframe (also known as the Vendian) is when the first fossils of morphologically complex organisms arrive on the scene. (Molecular clocks suggest that metazoans (animals) may have evolved as much as 1.2 bya, but there is no fossil evidence of them back anywhere near that far. Problems with molecular clock calibration between vertebrates and invertebrates, and the likelihood that mutation rates are a function of metabolism rather than time, make it probable that the molecular clock dates are too old by as much as a factor of two.) Named for the Ediacara Hills in South Australia where the first well-preserved assemblage of these fossils was discovered, the Ediacaran biota has been an enigma from the start. What does this assemblage of strange, soft-bodied creatures represent? Are they early animals? Complex photosynthesizers, whether independent (like plants) or containing symbionts (like corals)? Something else altogether, evolutionary dead-ends? There is no end of suggestions.

The fossils themselves resist attempts to "shoe-horn" them into extant phyla. Some vaguely resemble sea pens; others jellyfish; most nothing at all—a strange assortment of discs, blobs, spindles, lumps, and other nondescript shapes. Not a one of them has hard parts—those wouldn't evolve until the Cambrian. Some seem to have holdfasts and thus were sessile, while others may have been able to move around. What is clear, though, is that they must have been multicellular to grow so large (some as much as a meter). Truly complex life had arrived on the scene.

Toward the end of the Ediacaran, the first trace fossils—clear evidence of animals moving around—appear in the rocks. The oldest horizontal burrows are about 575 million years old, but vertical burrows do not appear until the start of the Cambrian.

Where Precambrian Rocks Are Found

In large part, if you want to locate Precambrian rocks in North America, you should look to the Canadian Shield. Being the old, stable core of the continent, and recently having been bulldozed clean by glaciers, much of the bedrock in eastern Canada, Greenland, and the Arctic Archipelago is Precambrian in age. Turn to the inside cover, and look at the map on the endsheets. The reddish rock areas are Precambrian. Once glance at this map, and the Canadian Shield jumps right out at you.

Elsewhere on the continent, Precambrian rocks are exposed at the surface mainly in mountainous areas or deep gashes like the Grand Canyon. Many of these rocks formed when the terranes of which they were a part were exotic, i.e. not yet part of North America—but that makes them no less North American now. Our continent has a rich rock record from Precambrian times, but unfortunately for fossil hunters, few of these rocks are suitable for collecting—unless you have a laboratory setup to look for microfossils.

Tracks of unknown creature from Precambrian-Cambrian boundary sediments, *Climactichnites wilsoni*, Dresbach, Wisconsin. North American Museum of Ancient Life, Lehi, Utah.

Where Precambrian Fossils Are Found

That's what you really wanted to know, wasn't it? Not just where the Precambrian outcrops are, but where you can find the *fossils*. Obviously, the latter is a subset of the former—only sedimentary, not igneous or metamorphic, rocks will do—and if we exclude those pesky microfossils (as we'll largely do for the remainder of the book), it is a small subset indeed. It's impossible to explain in a couple of paragraphs where all the best localities are; I'll merely hit some of the highlights. Disclaimer: I make no promise—no, nary a hint—of easy access to some of these areas, particularly those in the Arctic! And it goes without saying that you should always obtain permission before prospecting on private lands.

As stromatolites are the most common Precambrian macrofossils, we'll start with those. (Zip up your parka.) Good stromatolites have been reported from the Eleonore Bay Group, Canning Land, East Greenland; from the Kuuvik Formation of Bathurst Inlet, Northwest Territories; from the Dismal Lakes and Rae Groups, District of MacKenzie, N.W.T.; in the Yellowknife Supergroup, Slave Province, N.W.T.; from the Borden Basin, Baffin Island, Canadian Arctic Archipelago; and from the Mavor Formation on Belcher Island in Hudson Bay. What? You sold your parka at a garage sale? Ok, let's see if we can't find some in warmer latitudes...like the Sibley Group of Thunder Bay, Ontario; the Michipicoten Group near Wawa, Ontario; or the Purcell Supergroup of the Purcell Mountains in Alberta and British Columbia. Stateside you might check out the Copper Harbor Conglomerate in Michigan's U.P., the Biwabik Iron Formation in northern Minnesota, the Nash Formation in the Medicine Bow Mountains of Wyoming, or even the Chuar Group deep in the Grand Canyon of northern Arizona. Real desert rats might try looking in the Beck Spring Dolomite in California's Mojave Desert, or for an adventure south of the border, the La Ciénega Formation in Sonora, Mexico.

Although for the most part, I'm going to ignore microfossils, one particular formation has to be mentioned. This is the 2.1-billion-year-old Gunflint Chert of the Thunder Bay region in Ontario. Some of the best, undisputed Precambrian microfossils to be identified as such—bacterial filaments, and strange star- and umbrella-shaped forms—come from these rocks, thus they are of uncommon import to the Precambrian fossil story.

For many, if not most, fossil aficionados, however, their real desire when it comes to the Precambrian is to uncover some of those wild and wonderful Ediacarans. The best North American assemblage found to date is in the Mistaken Point Formation of southeast Newfoundland (see below). Ediacaran critters have also been reported from the Windermere Supergroup in the Wernecke Mountains, Yukon Territory; from the Miette Group in the Rocky Mountains of British Columbia; from the Shaler Group on Victoria Island, Canadian Arctic Archipelago; from the Cid Formation of the Albemarle Group at Jacob's Creek Quarry, North Carolina; from the Wood Canyon Formation in the White-Inyo Mountains of southeast California; and finally, from the Clemente Formation in the Caborca region of Sonora, Mexico.

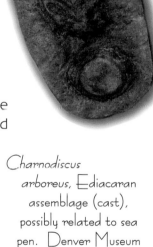

Charnodiscus arboreus, Ediacaran assemblage (cast), possibly related to sea pen. Denver Museum of Nature & Science, Denver, Colorado.

Where Can Precambrian Fossils Be Seen in the Field?

What if you're not a collector, or can't arrange permissions and logistics to prospect at your chosen Precambrian locality? There are still a couple of places where you can go and view the fossils, and armed with a good camera, come home with some nice mementos. Both of these sites require that you be in good enough shape to undertake some hiking, but you don't need to charter an icebreaker or a bush plane.

Without a doubt, the most spectacular Precambrian (Ediacaran) fossils known from North America are at Mistaken Point, Newfoundland, on the very southeast tip of the Avalon Peninsula. This is a protected reserve; you can drive to where there is a sign on the fence (ask the locals for directions—they are very proud of their natural heritage), but then you'll have to hike in several kilometers along a trail to where the fossils are exposed. One visitor said that the fossils were so plentiful that she was unable to avoid stepping on them as she walked around, much as she tried to be careful!

Stromatolites, 1.4 bya, Belt Supergroup, Glacier National Park, Montana. This has not been sectioned & polished; it is representative of what you'd see if you visited Glacier National Park. New Mexico Museum of Natural History, Albuquerque.

The other protected, and publicly accessible, place for viewing good Precambrian fossils is Glacier National Park in Montana. Several different rock formations there contain abundant stromatolites: the Altyn and Si-yeh limestones, and also the Appekunny, Grinnell, Snowslip, and Mt. Shields formations. An extensive network of hiking trails leads to many exposed outcrops. The best thing to do is to ask for trail recommendations at the visitor's center, telling them what you are looking for. That way you can select a hike in keeping with your stamina and interest level.

The boundary between the Precambrian and the Cambrian closely coincides with the appearance of a hard, shelly fauna, currently dated at 542 million years ago. Once animals evolved hard parts, evolution kicked into high gear, and the peaceful world of the Precambrian was gone forever. In the next chapter, we'll look at what happened during the Cambrian 'explosion.'

The Paleozoic Era

Life began in the water, and only slowly took to the land and air. A lobe-finned fish makes a foray onto terra firma in this mid-Paleozoic swamp in Québec.

ow that we've covered the first four billion years, we launch into the ~10% of Earth's history that to most of us is the most interesting—the three Eras of the Phanerozoic, when complex, multicellular plants and animals came to be common in the seas, and later took to the land and air. These three Eras are the Paleozoic ("era of ancient life"), Mesozoic ("era of middle life"), and Cenozoic ("era of recent life"), as shown on the geological time scale near the beginning of the book. These divisions of geologic time were first used by Giovanni Arduino in the 18th century; he divided the rock record into three slices, and originally used the terms Primary, Secondary, and Tertiary for them. A fourth division, the Quaternary, was added by Jules Desnoyers in 1829. The latter two names are still in use, but British geologist John Phillips renamed the Primary as the Paleozoic, and the Secondary as the Mesozoic, in 1860. The Tertiary and Quaternary were given the status of Periods, which he grouped together into the Cenozoic Era. Each of these Eras will form one of the three major sections of the remainder of our story.

The first of these is the Paleozoic Era, which runs from the base of the Cambrian at 542 million years ago to the end of the Permian around 251 mya. In between it encompasses the Ordovician, Silurian, Devonian, Mississippian, and Pennsylvanian periods (the latter two of which are lumped together in Europe and much of the world as the Carboniferous, but which are quite distinct in North America and will thus each be given a separate chapter). In examining the major paleontological developments of the Paleozoic, we will see the evolution of all the major phyla of animals which have good fossil records, including our own; we'll watch as invertebrates explode onto the scene and create the first reefs on our continental shelves; and we'll witness fish become top predators in the seas, and later crawl out onto land to become the first amphibians and reptiles. Of the four times that flight has evolved in the animal kingdom, we'll encounter it first among the insects. And we'll watch as plants, too, colonize the land, develop a vascular system, and become trees with seeds.

Let's go!

Chapter Two
The Cambrian

Anomalocaris attacks *Canadaspis* at the foot of the Cathedral Reef in British Columbia 505 million years ago. Trilobites crawl through the limey mud on the bottom, while a school of lace crabs, *Marrella*, swims past a forest of sponges. Slumping of the reef, evidenced by the large chunks of debris at its base, would preserve these animals for us to find as fossils half a billion years later.

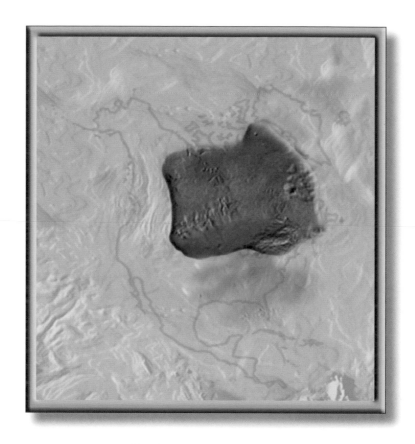

Cambrian North America

We launch now into the history of multicellular life—that of metazoans (animals), as we embark on our journey through the Paleozoic Era ("era of ancient life"). We'll come to plants later; at this point, they consisted of naught but algae, and so far as the fossil record is concerned, pretty much would until many millions of years later. The story of the Cambrian is the story of animals. It is the story of animals in the sea, as life would not invade land until much later. The continents—including ours—were barren, but the marine realm was teeming with strange creatures.

Definition and Nomenclature

The Cambrian Period was named in 1835 by the British geologist Adam Sedgwick. He derived the moniker from Cambria (or Cumbria) in northern Wales, which was the Latin name for the region—and where the first recognized rocks of this age were found. The term has its etymology in the Welsh word "cymry," meaning "countryman" or "compatriot" (against Anglo-Saxon invaders).

As we saw in examining the Precambrian, evolution had been busy for billions of years before the first fossils of multicellular creatures are recorded in the sediments. Although in 1991, the International Subcommission on Cambrian Stratigraphy formalized the Precambrian-Cambrian boundary at the first appearance of the horizontal burrow *Trichophycus pedum* in the reference section at Fortune

Head, Newfoundland (age 542 mya), this approximately corresponds with the appearance of a widespread, diverse fauna possessing easily fossilizable hard parts (the traditional beginning of the Cambrian as recognized by early geologists). Scientists as far back as the nineteenth century, including Charles Darwin, understood that something special marked the onset of the Cambrian; life, in all its fabulous diversity, seemed to spring almost magically and instantaneously from the vast void of the Precambrian.

Until the advent of radiometric dating, there was no way of knowing how long ago that was, or how much time the Cambrian Period represented. The earliest radiometric dates—dates derived from the decay of radioactive minerals in rocks—indicated that the Cambrian began shortly after 600 mya and spanned almost 100 million years. More accurate dates obtained in the last two decades have resulted in a revision of the chronology and a compression of the length of time the Cambrian Period represents. The overall timeframe has been shifted towards the recent in the process. In 1998 the Cambrian-Ordovician Boundary Working Group adopted a biostratigraphical definition of the beginning of the Ordovician (which is the same as the end of the Cambrian)—the first appearance of the conodont *Iapetognathus fluctivagus* in the rocks at Green Point, Newfoundland—with an age of 488 mya. Thus the current definition of the Cambrian encompasses an interval of 54 million years.

Trilobite *Athabashia wasatchensis*, Spence Shale, Antimony Canyon, Utah. Private collection.

Paleogeography and Paleoclimate

Let's orient ourselves with an overview of Cambrian paleogeography and climate. We left the continents in the late Precambrian mostly assembled into the supercontinent of Pannotia, a reincarnation of Rodinia. But North America, welded to Greenland and portions of what are now Great Britain, France, and Scandinavia, drifted alone. By the onset of the Cambrian, our continent had moved north from its previous position near the South Pole into more equable climes. It occupied the midlatitudes of the southern hemisphere and was oriented about 90° clockwise from its position today; thus what is now the Pacific Coast of North America lay approximately parallel to the Tropic of Capricorn, and Greenland and Québec extended to around 50° south. Sea levels were high and much of North America, especially the lands which comprise the southern and western United States today, lay underwater. Parts of modern Newfoundland, Nova Scotia, and New England had not yet docked with the remainder of North America, instead forming a submarine portion of the small continent of Avalonia, which was located near the South Pole. Calm, shallow seas occupied shelf environments marginal to the Canadian Shield, and were ideal for the formation of extensive limestones and reef deposits. Slumping and mudslides along the scarp of the continental shelves formed many of the most spectacular Cambrian fossil deposits found worldwide to date. These include British Columbia's Burgess Shale, the Kinzers Formation of Pennsylvania, and the Buen Formation of Peary Land, northern Greenland, from which the Sirius Passet fauna derives. All of these fossil deposits preserve amazing animal assemblages that include a host of soft-bodied organisms in addition to those with hard shells. Although other such Cambrian lagerstätten (a German mining term meaning 'mother lode,' applied to particularly spectacular fossil localities) are known—most notably the Chengjiang fauna from China—we have more than our fair share of them, because North America had relatively more coastline than did the other continents (since they were agglomerated together). Pannotia broke apart during Cambrian times, but that need not concern us here, as it is not part of North America's story.

Trilobite *Modocia laevanuca*, part & counterpart, Marjum Shale, House Range, Utah. Private collection.

Climate during the Cambrian is not well known, although there are no indications of any glaciations occurring during this time. Without large continental landmasses located in the high latitudes of either hemisphere, oceanic currents would have been able to circulate freely and distribute heat relatively well between the equator and the poles. Thus, it is likely that worldwide climates were fairly equable, being neither overly hot nor terribly cold. Early in the Cambrian there is evidence of aridity over much of the North American continent—consistent with its positioning underneath the southern subtropical dry-weather belt. This is created by the persistent high pressure of the convergence zone between the equatorial and midlatitude atmospheric circulation cells. (These convergence zones, one in the northern hemisphere and one in the southern at about 30° latitude, are responsible for the presence of many of the deserts we have today—most notably the Sahara and the Australian Outback.) Later on in the Cambrian, North America had drifted further northwards such that Alaska straddled the equator. Tropical climates probably existed along the modern western (ancient northern) margin of the continent, and warm-temperate ones along the modern eastern (ancient southern) shore.

The Small Shelly Fauna

The first mineralized fossils to be found in Cambrian sediments occur a little above the basalmost trace fossils and have been named the "small shelly fauna." They consist of a diversity of small spines, plates, tubes, and cones that are rarely found with the remains of the animals that sported them. Enigmatic for decades, recent research has shown them to be body ornamentation borne by small, worm-like organisms. Perhaps they were the first attempts at defensive armature against the evolution of predatory animals.

The Cambrian 'Explosion'

No, it's not a massive volcanic eruption or huge asteroid impact; this is a term popularized in the 1970s for an event that was known to scholars more than a hundred years before—the seemingly sudden appearance, in Cambrian rocks, of a huge diversity of animals with disparate body plans. In fact, representatives of all the phyla with easily-fossilizable hard parts that are known from the fossil record appear in the Cambrian, with the exception of sponges (present in the latest Precambrian) and bryozoa (which do not appear until the subsequent period, the Ordovician). We can only speculate whether representatives of modern, solely soft-bodied phyla without fossil records (like many worms) may have entered the scene during Cambrian times as well.

Trilobite *Modocia brevispina*, Marjum Formation, House Range, Utah. Note the predator bite mark. The fact that the rest of the trilobite's carapace is intact suggests that this one got away.
Private collection.

Charles Darwin pondered this sudden appearance of so many animals, and attributed it to large imperfections in the fossil record. Certainly, such gaps exist, but another century and a half of fossil collecting has failed to remove the apparent suddenness of the 'explosion.' We have seen previously that during the Ediacaran (latest Precambrian), there is ample fossil evidence of a diverse and abundant array of soft-bodied lifeforms in the seas. These, however, were preserved only under the most unusual of circumstances because of their lack of hard parts. Furthermore, few Ediacarans resemble Cambrian animals, or are even assignable to extant phyla. Arthropods (especially trilobites, but also crustaceans and a host of weird forms), molluscs, echinoderms, corals, annelids (segmented worms), brachiopods, graptolites, tunicates, and chordates are all represented in Cambrian sediments. The aforementioned lager-stätten also reveal a plethora of strange creatures neither certainly belonging to any extant phylum nor anything like the Ediacarans, either. The $64,000 question surrounding the 'explosion' is this: did these meta-zoans spring *de novo* from the abyss in the early Cambrian, or did something in the environment change, merely allowing their fossils to first be preserved at this time? In other words, was there really an explosion of evolution, or is the sudden increase in animal fossils a preservational artifact?

One fairly robust piece of data indicating that there was nowhere near the diversity of metazoans during the Late Precambrian as there was later is the paucity of trace fossils found in the older sediments. Had there been a diversity of vagile (movement-capable) animals, even soft-bodied ones, one would expect to find an abundant and varied trace fossil record during the Late Precambrian. This is not the case. Horizontal burrows do not occur in sediments older than about 575 mya, and vertical burrows before 542 mya. This is clear evidence that many econiches were not filled during the latest Precambrian, and circumstantial evidence that the basic types of organisms—possibly the phyla themselves—capable of exploiting those niches had not yet evolved.

Let us begin with the assumption that the 'explosion' is real—that all the modern phyla with readily-fossilizable hard parts, i.e. all those with significant pres-ervation potential, did indeed evolve during the 54 million years of the Cambrian (and, the best evidence suggests, within a 10-million-year interval between 530 and 520 mya). Not surprisingly, sediment burrowing be-comes more common and extensive throughout the Cambrian—just what we'd expect if diversity was, in fact, increasing. What might have driven such a burst of evolution, and why has the pace slowed since?

Many scenarios have been proposed over the years, some more and some less ro-bust in the face of scrutiny. For example, the breakup of Pannotia would have resulted in an increase in continental shelf area and a corre-sponding increase in habitat for benthic and shal-low-water marine creatures. More habitat, more econiches, more animals to occupy them. But then, why wasn't a burst of evolution associ-ated with the fracturing of Rodinia? Either it was—culminating in the Ediacaran biota some

Carpoid *Castericystis vali*, Marjum Formation, House Range, Utah. Private collection.

50 million years later—or, if not, this may have been because, back then, only uni-cellular progenitors existed, so evolution lacked the "raw material" from which to fashion the elaborate body plans that it could during the Cambrian. Perhaps, evolu-tion was spurred on to refill econiches that were vacated by extinctions related to the late Precambrian glaciations. Both of these suggestions have merit, but similar

circumstances occurred later on in the geologic record (e.g., subsequent to the end-Permian extinctions), yet they were accompanied by no evolution of new phylum-level body plans. Probably these were, at most, contributing factors...and we must look elsewhere for the primary cause(s).

Another intriguing suggestion is that the Cambrian Explosion corresponds with the invention of sexual reproduction. (We have mentioned this in the last chapter, but no one is really sure just when it occurred.) Evolution suddenly could create variety far more quickly and easily than it could through mutation alone. This certainly sounds rational, but is difficult, if not impossible, to test. Another plausible suggestion is that the rate of evolution was accelerated by an "arms race" between predators and prey that had not existed during the peaceful days of the Precambrian, when presumably most, if not all, animals were sessile filter-feeders. We have noted that trace fossils—which indicate vagile animals—are rare prior to the Cambrian, which lends some credence to this possibility. Also, Ediacaran fossils showing clear evidence of defensive adaptations are unknown, whereas the Burgess Shale is full of critters sporting spines, horns, and other structures which could have been used for protection. Obvious predators, notably anomalocarids and conodonts, abound. Arthropods in the Burgess Shale are particularly diverse, and show that trilobites, while the dominant fossils of the Cambrian, were merely a minor branch of the arthropod family tree at the time (one which did possess mineralized carapaces—a possible reason for their success). In fact, the finest lesson of the Burgess Shale may be to show us how incomplete and skewed the Cambrian fossil record really is. Perhaps the hard parts which exponentially increased the preservation potential of Cambrian organisms were a necessary defensive response to the evolution of active predators.

Trilobite *Modocia typicalis*, Marjum Formation, House Range, Utah. Private collection.

What about the proposition that the Cambrian Explosion is an illusion—that complex metazoan life existed far back into the Precambrian, and is only revealed to us by fossils when hard parts, and greatly increased preservation potential, evolved? Reason to suspect this comes from recent studies of genetic divergence and the idea of a 'molecular clock,' again something we briefly mentioned, but did not delve into, in chapter 1. A molecular clock is based on the idea that mutations in an organism's DNA occur at a statistically constant rate throughout time (although this is far from proven, and is merely a reasonable assumption). If true, the number of mutations separating the DNA of two organisms—or genetic difference between them—provides an estimate of the time since they diverged from a common ancestor. Such a 'clock' must be calibrated against fossil data, and is only accurate if the mutation rate truly has remained constant throughout time; therefore, it can only give ballpark estimates, but these are useful nonetheless.

It has been pointed out that such studies reveal that evolution at the molecular (genetic) level frequently proceeds much more quickly than at the level of morphology (which is all the fossil record can reveal). Two organisms that differ greatly in their genetic sequence may not appear very far apart morphologically. Prior to genetic sequencing, Bacteria and Archea were classified together because of their strong physical resemblance. DNA sequencing reveals that they differ from each other as much as do ostriches and liverworts—and therefore obviously have been on separate evolutionary paths for a very long time. Fossil evidence cannot reveal this disparate history. Proponents of this point of view point out that molecular studies place the common ancestors of the modern phyla deep within the Precambrian, some billion or more years ago. That supports the "illusion" theory of the explosion.

Cambrian stromatolites, Box Elder Co., Utah. Denver Museum of Nature & Science, Denver, Colorado.

However, I have my doubts about this argument. The reverse is often true: a small difference in DNA can result in a paleontologically significant difference in morphology. We share 98% of our DNA with chimpanzees, yet no primate paleontologist would mistake a chimp for a human in the fossil record. This is especially true when the mutations occur in homeobox (Hox) genes—those which control the timing and trajectory of developmental processes (rather than the synthesis of proteins). A minor mutation in a developmental gene can result in as significant a change in morphology as a bird growing teeth or not. (It has even been argued that mutations in Hox genes fuel the "jumps" in punctuated equilibrium.) Additionally, I find the assumption that genetic mutations accumulate at a constant rate, especially over the expanses of geologic time and across diverse lineages, rather tenuous. A small difference in the rate of mutation between two lineages can have a tremendous effect on the divergence dates calculated. (In fact, mutation rates derived from vertebrate animals are often extrapolated to invertebrate lineages—a big jump, and one that is not necessarily justified.) This is not to say that I do not believe that various phyla diverged from one

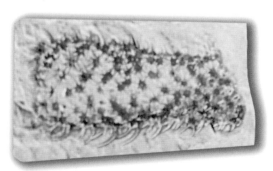

Sponge *Chancelloria pentacta*, Wheeler Shale, House Range, Utah. Private collection.

another far back in the Precambrian. They well may have—but I am skeptical of putting too much faith in molecular clock studies as evidence of it.

A frequently-cited argument for the illusory nature of the 'explosion' is that the Cambrian environment was somehow different from that of the Precambrian, in ways that allowed animals to develop hard (fossilizable) shells when they were unable to before. The two most common factors proposed are an increase in dissolved calcium in the seas (from rock weathering) and a newfound abundance of oxygen (from photosynthesis). Either of these, or both together, would make synthesis of calcium carbonate—the main material these animals used in their shells—much easier than it had been when such levels were low. This is a reasonable suggestion, but evidence for higher calcium levels is lacking, and oxygen was increasing only very gradually at this time, as there were no new photosynthesizers present that had not already existed during the Precambrian. Again, these may be contributing factors, but probably not the whole story.

So what can we conclude from all of this? Taken together, there is not a clear-cut case for either a wide radiation of phylum-level body plans (archetypes) deep within the Precambrian, nor for the entirety of the Cambrian Explosion being explained by "evolution in overdrive." Like so many things in nature, it needn't be all one or the other. Probably many lineages had embarked on widely disparate trajectories long before the Cambrian began, yet evolution itself also spun into high gear during the interval from 530 to 520 mya. One explanation for this which I find to be quite convincing is that the timespan encompassing the Cambrian Explosion represents the maximum slope of a sigmoid growth curve—the portion subsequent to the slow-increase, "get off the ground" beginning and the asymptotic amelioration that sets in when econiches become filled or limits on food supply make themselves felt.

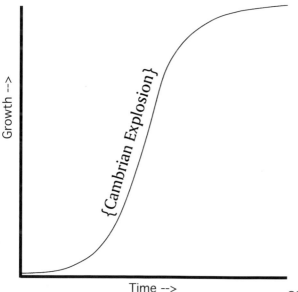

Say what?—puzzle you non-mathematical types! Ok, a little explanation is called for. Take a look at the accompanying graph. It is shaped more or less like a forward-slanting S, flat on the top and the bottom (at the beginning and ending of the stroke) and almost vertical, rather than backwardly diagonal, in the middle. Height is population or diversity (could be number of individuals, number of species, or number of archetypes) and increasing distance to the right represents time. Say you have bacteria dividing in a petri dish (or organisms giving rise to new species). Say every minute (or million years) they double—give rise to two daughter cells (or species). Each of those progeny doubles during the next unit of time, etc. etc. Try it on a calculator, starting with 1 and multiplying by 2 each time.

You'll have a million in less time than you think. But this can't go on forever—those organisms need resources (food, living space) and those resources are finite. So at some point, some of them begin to starve (or die off for whatever reason), and you don't have the entire lot of them doubling their population with each iteration. The longer you go, the more of them can't compete, and the slower the growth rate gets until it's effectively zero and the population is limited by the carrying capacity of the environment; you are at the top of the S—time goes on, but your population (or number of new species) fails to increase with it.

Arthropod *Canadaspis* sp., Burgess Shale, British Columbia. Smithsonian travelling exhibit on display at Sternberg Museum of Natural History, Hays, Kansas.

This may have been what happened during the Cambrian. New taxa evolved at an ever-increasing rate until the limits of ecospace began to impinge on their partying. Why have no new archetypes or phylum-level body plans appeared subsequent to this interval (with the exception of the bryozoa, which didn't lag by much, and might have been present in soft-bodied form anyway)? Perhaps eco/morphospace was filled and no one new could squeeze in. Perhaps the necessary early-stage mutations were all lethal. Maybe genetic baggage gave animals a certain amount of morphological 'inertia.' Who knows. Suffice it to say that by the end of the Cambrian, the blueprint for the next half billion years of life on Earth had been laid.

Life in the Cambrian

Ok, enough about the 'explosion.' What was a typical Cambrian ecosystem like? We've noted that it was, so far as we know, entirely marine, and that it lacked non-algal plants and swarmed with disparate and strange animals. It used to be thought that the soft-bodied Ediacaran biota had become extinct by the dawn of the Cambrian. Recent discoveries, such as one in the Salt Spring Hills of California, have debunked this notion and proven that some Ediacaran critters survived at least into the Early Cambrian. In large part, however, these holdovers were replaced by new animals that appeared during the 'explosion.'

We have mentioned that there were reefs fringing the shores of the Canadian Shield. These reefs, however, weren't formed by corals—they were created by skeletonized sponges

Trilobite *Wanneria* sp., Eager Formation, Cranbrook, British Columbia. Private collection.

called archaeocyathids, which had conical, cup-like shapes and grew in dense colonies. Calcareous algae also helped cement the reef together. And these reefs were home to a plethora of animals quite unlike anything we know today.

The Cambrian is known to many as the "Age of Trilobites" because trilobites are the most common fossils found in Cambrian rocks. But Sirius Passet, the Burgess Shale, and other lagerstätten teach us that whereas yes, arthropods were the dominant phylum, trilobites themselves were relegated to a minor role compared with non-trilobite arthropods. It would have been amazing enough to be transported back to a trilobite reef, but one populated by the bizarre variety of arthropods revealed in the Burgess Shale would have been even more awesome. Although the other modern phyla were present, their numbers and ecological importance paled in comparison with those of the arthropods. Anomalocarids—considered by some to be arthropods, by others to be a separate phylum—preyed on the trilobites and their kin. Brachiopods, other landmark Paleozoic denizens, consisted at this time of mostly inarticulate forms. Molluscs were rare; hyolithids (an extinct group), bivalves, and gastropods were present, but the cephalopods were the first molluscs to gain significant ecological importance. Echinoderms consisted of mostly now-extinct forms such as edrioasteroids and helicoplacoids; a few crinoids were present, but the familiar starfish and sea urchins had not yet evolved. Jawless chordates (representative of our own phylum, and including forms such as the enigmatic conodonts and the torpedo-shaped *Pikaia*) swam through shallow waters. Some of the Chengjiang fossils have been interpreted as soft-bodied fish, but in North America, fish would not appear until the Ordovician.

Trilobite cf. *Niobella* sp., Random Island, Newfoundland. Private collection.

The Cambrian came to a close with a mass extinction that decimated 75% of the trilobite families, half the sponge families, and a significant portion of the brachiopods and snails. Archaeocyathid sponges had gone into severe decline during the middle of the period, and wouldn't survive the end. A possible reason for all of this was changing environments associated with the breakup of Pannotia, but in truth, it really remains a mystery.

Where Cambrian Rocks Are Found

Eocrinoid *Gogia spiralis*, Wheeler Shale, Utah. North American Museum of Ancient Life, Lehi, Utah.

Cambrian rocks in North America are mainly found in mountainous areas and in the Arctic. Northern

Greenland and the islands that make up the Canadian Arctic Archipelago have many extensive exposures. Some Cambrian rocks outcrop near Montréal and continue up along the St. Lawrence River, past Québec City and up the Gaspé Peninsula. Intermittent exposures are also to be found throughout Nova Scotia and Newfoundland. These continue southwest through the Appalachians all the way into Alabama. Some Cambrian exposures can also be found in southeastern Missouri. There is a large band of them across the midsection of Wisconsin.

The remainder of North America's Cambrian rocks occur in the Rocky Mountain West. Large exposures are found in the mountainous areas of the Northwest Territories and the Yukon, and in eastern British Columbia on into southwestern Alberta. This band continues down the central spine of the Rockies into Colorado and New Mexico in a very disjointed fashion. Cambrian sediments are extensively exposed in the Great Basin region of southeast Idaho, western Utah, across Nevada, and including eastern California. Isolated outliers crop out in Baja California and northwestern Sonora, Mexico, as well as some small areas in the state of Guerrero, but there isn't much Cambrian rock south of the border.

Where Cambrian Fossils Are Found

Trilobites are the quintessential Cambrian fossils, and that is what most collectors are after when they hunt for critters in rocks of this age. In the eastern areas of the continent, collecting opportunities are limited, but Alabama and Georgia's Conasauga Formation has been known to yield spectacular agnostid trilobites in isolated pockets, as well as the strange "star cobble" *Brooksella* (which has been recently reinterpreted as a type of sponge). Pennsylvania's Kinzers Formation has already been mentioned for soft-bodied fossils; it contains many trilobites as well. The Hanford Brook Formation of New Brunswick is a classic unit for well-preserved trilobites. Outcrops of the Brigus Formation in Newfoundland, Nova Scotia, and Massachusetts are sometimes quite fossiliferous. Trace fossils called *Diplichnites*, possibly made by trilobites or other arthropods, have been found in Wisconsin's Mt. Simon Sandstone, along with other trace fossils *(Climactichnites)* of enigmatic origin. The latter also occur in Missouri's Lamotte Sandstone. Above the Lamotte, the Bonneterre and Eminence formations are good places to hunt for trilobites, brachiopods, and monoplacophoran molluscs.

Growth series of the trilobite *Elrathia kingi*, the most common species found in the Wheeler Shale. Antelope Spring, Utah. Author's collection.

Most trilobite collectors, however, head to the basin and range country of Utah and Nevada, where countless small fault-block mountain ranges have exposed Cambrian sediments. Undoubtedly the most famous of these is the House Range in west-central Utah, where the Wheeler Shale, the Marjum Formation, and the Weeks Formation literally teem with body fossils and molts of a number of trilobite species. These formations also yield small inarticulate brachiopods, and sometimes sponges. The Pierson Cove Formation in the nearby Drum Mountains has fewer fossils but is still a good place to look for trilobites and hyolithids. The Spence Shale, St. Charles Limestone, and Wilbert Formation in Idaho contain trilobites and brachiopods. The Chisolm Shale near Pioche, Nevada has produced some spectacular trilobites, and the nearby Pioche Formation is reported to contain a Burgess Shale-type soft-bodied fauna.

Trilobite Elrathia kingi, Wheeler Shale, Antelope Spring, Utah. This species is frequently found weathered free from matrix, just waiting to be picked up. Author's collection.

California's Wood Canyon Formation is generally not very fossiliferous, but some sites are reported to produce scattered olenellid trilobites and some archaeocyathids, in the upper part of the formation. The real action in the Mojave Desert is the Latham Shale, exposed in the Marble and Bristol Mountains. This is a classic trilobite unit well-known to many collectors. Just above it, the Chambless Formation is rich in oncolites, which, like stromatolites, are the remains of cyanobacteria, but differ in that they form small, spherical growths instead of large columns. The next-youngest unit, the Cadiz Formation, contains some trilobites and many trace fossils. Elsewhere in the Mojave, these three formations are grouped together as the Carrara Formation.

Where Can Cambrian Fossils Be Seen in the Field?

Sponges Chancelloria pentacta, Wheeler Shale, Utah. North American Museum of Ancient Life, Lehi, Utah.

No question about it—the premier locality for North American Cambrian fossils is British Columbia's Burgess Shale, a part of the Stephen Formation. First explored by Smithsonian paleontologist Charles D. Walcott in 1909, his original quarry is still being worked. Discoveries made at this site, in Yoho National Park near Field in the Canadian Rockies, demonstrated that trilobites owe their dominance as fossils to their mineralized carapaces—while in fact, they made up only around 10% of the arthropods in this Cambrian fauna. Rapid burial by mudslides at the reef front allows the Burgess Shale to present an unbiased snapshot of a Middle Cambrian (505 mya) community.

Walcott's quarry can be visited by those capable of undertaking a several kilometer hike with steep vertical gain, but only with a registered guide—you cannot go on your own, and you cannot collect any fossils. Nevertheless, this world-famous lagerstätte is well worth the trip if you have the time and physical ability. Its importance is recognized by its designation as a World Heritage Site. If you go, be sure to keep your eyes peeled for the lace crab *Marrella*, the most common (and perhaps the most beautiful) fossil in the quarry.

Eocrinoid *Gogia* sp., Spence Shale, Wellesville Mountains, Utah. Private collection.

Another public preserve where Cambrian fossils may be seen *in situ* is far to the south, in Death Valley National Park, California. The Paleozoic geology of Death Valley is amazingly complete, and we'll mention it several times as a place where you can see fossils from this timeframe, if you're willing to do some hiking. The difficulty here is not so much mountain-climbing as being prepared for desert conditions (go in the wintertime—the valley didn't get its name without reason). Be sure you are fit, well-equipped, and have plenty of water, and never hike alone. Check with a ranger before heading out into the backcountry, so as to be aware of any restrictions, and to be sure someone knows where you're going. The fossiliferous Cambrian units which outcrop in Death Valley are the Nopah and Carrara formations, both of which contain many fragmented trilobites, as well as gastropods in the former. As with all national parks, take only pictures home with you.

And finally, if you're in the Northeast, you can see Cambrian stromatolites preserved at the Petrified Gardens near Saratoga Springs, New York. These algal mounds are in the Hoyt Limestone and are considered the best example of Cambrian stromatolites in eastern North America.

Trilobite *Asaphiscus wheeleri* (approximately life size), Wheeler Shale, Antelope Spring, Utah. The author's husband overlooked this one on their honeymoon so that she could find it. Author's collection.

————————————————

The Cambrian was an instrumental time for life on our planet. When it dawned, only soft-bodied creatures swam and crawled through the seas; by its end, complex ecosystems had evolved, and the modern phyla were in place. It was a time of experimentation among animals, both ecologically and morphologically; not all the players would survive. Much of subsequent evolution was constrained to follow the paths which were paved during the Cambrian. The stage had been set for the next half-billion years of the saga of life on Earth.

Chapter Three
The Ordovician

The greatest diversification in the history of animal life took place during the Ordovician.
Trilobites crawl across a muddy seafloor in southern Ohio decorated with crinoids and horn corals, trying
to stay clear of the grasping tentacles of hungry nautiloid cephalopods.

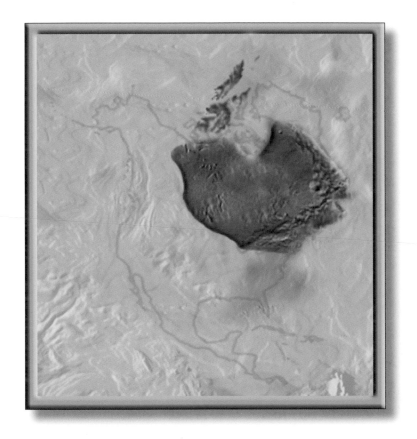

Ordovician North America

You might think, from all our talk in the last chapter about the Cambrian Explosion, that evolution took a much-deserved holiday during the immediately succeeding times. Nothing could be further from the truth. While the major animal phyla were in place by the close of the Cambrian, the Ordovician Period actually saw more new orders and families arise in these phyla than ever before. This has been called the "great Ordovician biodiversification event," and it would see new players emerge that would flourish throughout the Paleozoic.

Definition and Nomenclature

The Ordovician is the second in sequence of the time periods of the Paleozoic Era. The flanking periods, the Cambrian (preceding) and Silurian (following), were, in fact, defined first. In 1831, British geologists Adam Sedgwick and Sir Roderick Murchison began working on the stratigraphy of northern Wales. Sedgwick began his work at the bottom of the sequence, and named his rocks the Cambrian. Murchison chose the name Silurian for the rocks he was examining toward the top of the same sequence. Sedgwick worked his way upward through the stratigraphic column, and Murchison worked downward from the top; eventually they met up, and both claimed the rocks in the middle for their own. For over 40 years this portion of the geologic column was in dispute until, in 1879, Charles Lapworth introduced the name Ordovician for the intervening sequence. The name was derived from that of an ancient Celtic tribe, the Ordovices, who once inhabited this region of Wales. The new name took a while to catch on, though, and only gained widespread support in the middle of the twentieth century. It was officially sanctioned for global use by the International Geological Congress in Copenhagen in 1960.

In 2000, the Cambrian-Ordovician Boundary Working Group adopted a biostratigraphical definition of that boundary—the first appearance of the conodont *lapetognathus fluctivagus* within bed 23 at the 101.8 m level at Green Point, Newfoundland. (Conodonts, mentioned briefly in the last chapter, are the tiny tooth-like elements found in the pharyngeal region of some extinct soft-bodied chordates—members of our own phylum. They are widespread in Paleozoic rocks and thus very useful for biostratigraphy.) Current dating places the age of this horizon at 488 million years ago. Lapworth had located the Ordovician-Silurian boundary in an uninterrupted sequence of black shales exposed at Dob's Linn in southern Scotland. In 1984, this locality was chosen as the international stratotype for the transition and the boundary defined as the first appearance of the graptolites *Parakidograptus acuminatus* and *Akidograptus ascensus*, 1.6 m above the base of the Birkhill Shale, with a date of 444 mya. (Graptolites, again members of our own phylum, were colonial hemichordates.) Thus the Ordovician Period may be considered to encompass some 44 million years.

"Lace collar" trilobites *Cryptolithoides ulrichi*, Viola Limestone, Arbuckle Mountains, Murray Co., Oklahoma. Private collection.

Paleogeography and Paleoclimate

Where was North America at this time? Oceans stretched across the entire northern hemisphere north of the Tropic of Cancer. Our continent still occupied tropical latitudes, and was rotated some 45° or so clockwise from its present orientation. As it slowly drifted northward, the equator passed from Alaska into (what is now) western Canada. Early in the Ordovician, sea levels were high and global climates moist and mild. It was the time, in fact, of greatest cratonic submergence (highest amount of continental flooding) of the entire Paleozoic. "Superplume" activity—similar to a hotspot on steroids—has been invoked to explain this. (Hotspots are stationary regions of magma upwelling, such as those which are creating the Hawaiian island chain and the geyser activity of Yellowstone.) When upwelling activity is high, it creates a large volume of hot/warm crustal rock, which equals an increase in crustal volume. This "puffs up" the seafloor, and the ocean waters spill onto the continents.

By this time, the oxygen concentration in the atmosphere had reached about half what it is today. There was significantly more carbon dioxide (CO_2) than there is now, by a factor of between 8 and 18x (estimates vary). This led to a large greenhouse effect which was likely responsible for the warm climate. Much of North America was flooded with epicontinental seas that were perfect for the development of reefs made of stromatoporoid sponges, calcareous algae, and some early corals, which left their remains in extensive deposits of limestone. Significant portions of the Canadian Shield region of the continent, however, remained above water at this time.

Representative Ordovician fossils:

Although Pannotia had fractured during the Cambrian, many of the remaining large landmasses—with the notable exceptions of North America, Baltica (the core of modern Europe), and Siberia—were welded together into the supercontinent of Gondwana, which was comprised mainly of what are the southern-hemisphere continents of the modern world. As the Ordovician progressed, Gondwana drifted southward and Saharan Africa came to straddle the South Pole. Fossils from the seas surrounding these lands, including portions of present-day Nova Scotia, indicate that there was a strong latitudinal gradient to the fauna throughout the Ordovician—with, as today, high-diversity reefs occupying primarily tropical seas, and a relatively depauperate fauna at the higher southern latitudes. Slowly, Baltica approached North America from the east, and, as the two continents neared each other and the intervening Iapetus Ocean shrank, the edge of North America collided with several island arcs and microcontinents in what is known as the Taconian Orogeny. This tectonic activity thrust up high mountains (the 'Taconic Appalachians') stretching from New Jersey and Newfoundland through Greenland into Scotland and Norway. Although these were mainly folded mountains, there is evidence of volcanic activity toward the southern end of the chain.

above — brachiopods, *Rafinesquina alternata*, Waynesville Formation, Ohio; below — nautiloid cephalopod shell, *Endoceras* sp. Denver Museum of Nature & Science, Denver, Colorado.

The minicontinent of Avalonia, containing parts of the Canadian maritime provinces, rifted free of Gondwana and began the long journey north to meet us. Florida was wedged in where South America and Africa were joined. The whereabouts of most of the Mexican terranes are not well-known, but it is thought that some of them may have been joined to South America off Peru.

Trilobite *Encrinuroides capitonis*, Bromide Formation, Criner Hills, Oklahoma. Private collection.

Climates during the first half of the Ordovician differed significantly from those later in the period. As Gondwanaland settled on the South Pole, oceanic circulation and poleward heat transport were disrupted. There was a general lowering of the previously-high sea levels which, with the initiation of glaciation on the southern continent, produced a positive-feedback loop. The lowered sea level decreased the maritime influence in the interior of Gondwana, leading to cooling and the initiation of glaciers; increased albedo of the icecaps (reflection) cooled the climate further; glaciers in turn locked up waters that led to a further lowering of sea level...and so on. Recent computer simulations have suggested that continental drift alone is unlikely to have initiated the glaciations, however. Lowered CO_2 levels in the atmosphere may also have played a part. It has been suggested that the extensive reef development earlier in the Ordovician may have lessened the amount of atmospheric CO_2 by locking it up in carbonates (limestones). A slowing of seafloor spreading rates may also have provided the impetus for the initial drop in sea level via a decreased volume of rock at the midocean ridges. The factors are many and complex, and we may never fully unravel them. The result, however, was one of the coldest episodes in Earth's history, and the loss of a large fraction of shallow-water continental shelf habitat. This had devastating consequences for the fauna.

Trilobite *Isotelus latus*, Cobourg Formation, Bowmanville, Ontario. Note predator bite mark on pygidium; rounded edges indicate the trilobite has molted at least once since the attack, and thus survived. Private collection.

Although equatorial climates remained mild during the Ordovician Ice Age, the loss of habitat decimated tropical organisms as well as higher-latitude ones. To add insult to injury, when warming later occurred and sea levels rose again, they were accompanied by widespread anoxia (lack of oxygen) in the oceans. Corals were profoundly affected, primarily the tabulate types. A third of the brachiopod and bryozoan families, with more than half of the species, went extinct. A quarter of all invertebrate families (over a hundred of them) and 60% of the genera died out. Trilobites, graptolites, and conodonts were devastated. Overall, it was the second largest mass extinction ever (behind only the end-Permian one, which was yet to come).

Life in the Ordovician

So, if its designation was somewhat *ad hoc*, what events characterized the Ordovician and made it worthy of special designation? In fact, in the wake of the

Late Cambrian extinctions, many new families of invertebrates evolved and diversified in the Ordovician seas. Gone were the 'problematica' of the Cambrian—strange animals which resist pigeonholing into modern phyla. The anomalocarids, which had been the top predators in Cambrian seas, were replaced by the explosive diversification of nautiloid cephalopods. In part, this may have been a response to the increasing level of intelligence which cephalopods represented. With their large eyes and sensitive tentacles, cephalopods today (octopi, squids) represent the most intelligent invertebrate predators in the oceans. Whereas it is unlikely that Ordovician cephalopods approached modern ones in their level of braininess, they were likely a lot smarter than the anomalocarids which they replaced.

Receptaculitids (a type of colonial blue-green algae) *Ischadites iowensis*, Galena Formation, Illinois. Burpee Museum of Natural History, Rockford, Illinois.

Much evolutionary experimentation characterized the Ordovician, and the time represents one of the largest adaptive radiations in Earth's history. Ordovician communities were more ecologically complex than those of the Cambrian. Infaunal (burrowing) lifestyles were still uncommon—most organisms lived on the seafloor. The number of invertebrate families increased from around 200 at the end of the Cambrian to about 500 in the Early Ordovician. Foraminifera (marine amoebas with tiny shells) became widespread; plankton in general increased in numbers, and with them, there is seen a great increase in the number of filter-feeding organisms (in contrast to the predominant detritivores of the Cambrian). Both rugose (horn) and tabulate (colonial) corals appeared, and became prominent reef-builders. Bryozoans first appear in the fossil record and were the predominant colonial animals in Ordovician seas. Stromatolites, holdovers from the Precambrian, continued to decline in importance. Brachiopods diversified greatly, especially the articulate (hinged) forms.

Trilobites were still very numerous, although Ordovician types in general differed greatly from their Cambrian forebears. Many trilobites evolved elaborate horns, spines, and such, interpreted as defensive adaptations. Some trilobites developed eyes borne on stalks so that they could hide in the mud. Altogether they seem to have expanded the range of econiches they occupied. Trackway evidence indicates that a few trilobites entered brackish waters, pushing the frontiers of the marine realm for the first time. Trilobites are still very common among Ordovician fossils, though not quite as exclusively so as in the Cambrian. In fact, this was only the beginning of a long decline for the group, leaving just a few stragglers to become extinct at the close of the Paleozoic. One possible reason for both their commonness as fossils, and their presumed susceptibility to increasing levels of predation, is that trilobites did not resorb the calcium from their old carapaces prior to molting, as

Trilobite *Flexicalymene meeki*, Richmond Formation, Arnheim member, Mt. Orab, Ohio. Private collection.

modern crustaceans do. They left behind rather rigid, calcified exoskeletons with excellent preservation potential—many of which became fossilized, as evidenced by the frequency with which trilobites missing the 'free cheeks' are found (when you

encounter one like that, it was a molt, rather than a corpse). However, this lack of resorption of calcium left the newly-molted trilobite extremely soft and vulnerable to predation. In fact, even in modern arthropods, 80-90% of mortality occurs in association with molting; it's a scary time for a bug or a shrimp! (Spock, the Everglades blue crayfish I have in a small aquarium on my desk, always goes into hiding for a couple of days around molting time.) We've already mentioned the nautiloids as wily, effective predators, who no doubt fed on just about anything they could ensnare with their tentacles. They increased from one order at the start of the Ordovician to nine orders by its end. Our own forbears, too, were predatory in nature, and the small, soft-bodied fish found only rarely in the Cambrian fossil record were replaced in the Ordovician by more capable, carnivorous types—as we shall soon see.

Trilobite *Isotelus mafritzi*, Cobourg Formation, Colburne, Ontario. Private collection.

Crinoids, those quintessential echinoderms of the Paleozoic, radiated in large numbers and would decorate the seafloors throughout the Era. Early, odd echinoderm experiments died out, and starfish, sea urchins, and brittlestars all appeared. Most major bivalve lineages were established by the Middle Ordovician. Graptolites reached their zenith. A peculiar type of colonial blue-green algae, called receptaculitids, enjoyed tremendous success. These were plump discoidal colonies filled with holes that look a bit like pincushions. Conodonts were abundant, although fish were still uncommon. The first evidence of fishes in North America comes from the Harding Sandstone of Colorado and correlated formations. Most of these fossils are small, disarticulated pieces of body armor known as tesserae; they resemble the denticles found in modern sharks' skin (which make it rough and give it a use as sandpaper). However, the microstructure of these tesserae indicates that they are made of true bone and are undoubtedly from fish. One of the best-preserved early fishes, *Astraspis*, hails from Ordovician rocks in Colorado. All in all, twelve taxa of fish are known from our continent at this time, most of them agnathans (jawless fishes, like lampreys and hagfish). However, scrappy teeth and scales indicate that jawed (gnathostome) fishes were becoming established as well.

The face of the land, too, was changing. Finds of cryptospores indicate that simple, nonvascular plants such as mosses and liverworts boldly colonized low, permanently-wet terrestrial habitats. The fact that spores become less common in sediments with increasing distance offshore suggests that they have a terrestrial origin. No megafossils of land plants have yet been reported from the Ordovician, though. Lichens (a symbiosis of a fungus and an alga) may have been present as well, although the oldest undisputed fossil lichens come from the Devonian of Scotland. Rare finds of trilete spores hint that some of these earliest land plants may already have been vascular (with a water-plumbing system) and related to the rhyniophytes that we'll meet in the Silurian. There is even evidence that arthropods also attempted the conquest of the land. The oldest reported nonmarine trace fossils—caliche-encrusted soil burrows, probably from millipedes or related animals—come from central Pennsylvania's Juniata Formation. Trackways also attributed to millipedes are preserved in ancient coastal dunes of the Nepean Formation in Canada. For the first time, the continents sported a mantle of green—in some places, at least—to fracture the landscapes of bare, rocky crags and endless sandy flats, and the first animal pioneers began to investigate this new living space.

Where Ordovician Rocks Are Found

While Cambrian outcrops are sparse and scattered, Ordovician exposures in North America are much more extensive; we owe this good fortune to the sea-level highstand mentioned earlier, and to sediment deposition in the widespread, shallow seas. Again, the Canadian Arctic islands host some very large exposures, and there is a band of Ordovician sediments inland from, but concentric with, the southern shore of Hudson Bay. There are also outcrops along the northern and southern shores of Lake Winnipeg in Manitoba. The Great Lakes region is blessed with Ordovician rocks from southeastern Minnesota across Wisconsin and into the U.P. of Michigan, where they form the southern shore of Lake Superior eastward until they disappear beneath Lake Huron. This band crops out again along the southern shore of Georgian Bay, where it continues south and east to become the northern shore of Lake Ontario. Splitting in two, one band continues on up the Gaspé Peninsula of Québec

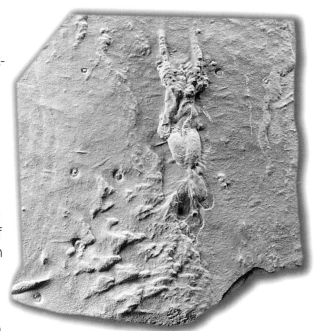

Trilobite trace fossil: Calymene tracks & resting place, Kope Formation, Cincinnati, Ohio. Private collection.

while the other cuts east across New York State. Some of the limestones of New York are among the best in the world in which to study the Ordovician. Outcrops occur along the trend of the Appalachians from Newfoundland to Alabama.

The best Ordovician exposures in the Midwest occur in Missouri and near the junction of Ohio, Indiana, and Kentucky, with another bit in central Tennessee. The Cincinnati area has a huge sequence of Late Ordovician sediments, and most of the southeastern third of Missouri is underlain by Ordovician rocks. Small outcrops are also found in the Black Hills region of South Dakota and Wyoming.

Again, the Rocky Mountains are host to a vast band of disjointed Ordovician outcrops beginning in Alaska and stretching into southwest Alberta. The basin and range province of southeast Idaho, western Utah, all of Nevada, and California east of the Sierras holds countless exposures.

Where Ordovician Fossils Are Found

If you want to rent a dogsled and "mush" all the way to Greenland for some millimeter-scale conodonts (they are, after all, some of our earliest relatives), check out the Cape Weber Formation in East Greenland. And the type locality for the beginning of the Ordovician—in the Green Point Formation of Newfoundland—was defined on the conodonts found there. Conodonts may not, however, thrill everyone because of their small size, so let's focus on macroscopic marine invertebrates.

Trilobite *Isotelus maximus*, Richmond Formation, Mt. Orab, Ohio. Some trilobites could roll up for protection; note also the long spines on the cephalon. Private collection.

The faunal list for Ontario's Verulam Formation is impressive: trilobites, brachiopods, bryozoans, clams, and snails, to mention a few. Abundant trace fossils are found in southern Ontario in the Georgian Bay Formation, and some trilobites have been reported from that province's Collingwood Shale. Amazing trilobites are found in the Trenton Group of New York, and trace fossils attributed to trilobites are found in the Clinton Group exposed there. "Beecher's Trilobite Bed," a widely-known New York collecting locality, occurs in the Frankfort Formation. This site is the most productive one known for trilobites with soft tissues preserved.

Ordovician fossils from the Midwest are nothing short of stunning. Uncounted spectacular trilobites have come from the Kope, Bull Fork, Fairview, Whitewater, and other Cincinnatian series formations of southern Ohio. The emergency spillway at Caesar Creek Reservoir exposes the Bull Fork Formation, from which come a lot of nice trilobites and crinoids, as well as bryozoans and other invertebrates. Indiana's Richmond Formation is a classic unit for brachiopods.

Kentucky's Bull Fork, Fairview, and Clays Ferry formations yield sponges, worms, bryozoans, brachiopods, trilobites, clams, snails, crinoids, and starfish. The Martinsburg Formation of Tennessee produces brachiopods, trilobites, bryozoans, and graptolites. The Nashville, Stones River, and Maysville groups of the same state contain brachiopods, bryozoans, and horn corals. Dolomites of the nearby Knox Group yield ostracods (tiny crustaceans with bivalved shells), clams, snails, and scrappy fish remains. Missouri's Gasconade Dolomite contains stromatolitic chert layers with abundant gastropod fossils (preserved in the act of grazing, perhaps?). The Roubidoux Formation of that state is highly fossiliferous and yields many cephalopods and snails, and the Plattin, Decorah, and Kimmswick Formations also are full of invertebrate fossils. Excellent trilobite fossils have come from the Galena and Maquoketa formations of the Green Bay region of Wisconsin. Corals and crinoids have been reported from southwest Wisconsin's Platteville Formation. Exposures of the Bromide Formation in Oklahoma sometimes produce myriads of trilobites. Abundant trace fossils and a few trilobites are found in South Dakota's Deadwood Formation.

Gastropods (snails) from Platteville Formation, Illinois: (top left & middle) - *Trochonema* sp.; (right) *Clathrospira conica*; (bottom right) - *Trochonema* sp.; (bottom left) - *Trochonema umbilicatum*. Burpee Museum of Natural History, Rockford, Illinois.

In the Great Basin, Ordovician outcrops bearing endocerid cephalopods and other invertebrates are exposed on Fossil Mountain in the Confusion Range of Utah. Most of these fossils are found in the Juab Limestone, the Kanosh Shale, and the Lehman Formation. The Pogonip Group of western Utah yields abundant, well-preserved invertebrates. Idaho's Fish Haven Dolomite is a good place to collect corals, brachiopods, crinoids, and snails. The Antelope Valley Limestone, exposed in the Toquima Range of Nevada, is chock-full of silicified trilobites, cystoid echinoderms, brachiopods, bryozoans, sponges, clams, and snails. Nearby, also in the Toquima Range, outcrops of the Vinini Formation yield wonderful examples of graptolites. The Great Beatty Mudmound is a large reef complex exposed about 180 km (110 miles) north of Las Vegas. It is well-known for its excellent trilobites, echinoderms, cephalopods, brachiopods, sponges, clams, and snails. The Johnson Spring Formation of California produces brachiopods, corals, and sponges.

Where Can Ordovician Fossils Be Seen in the Field?

There are quite a few preserves in North America that showcase Ordovician fossils. Possibly one of the most unique is at McBeth Point on the shores of Lake Winnipeg in Manitoba. Here is found a spectacular flora of macroalgae ("seaweeds"), along with some cephalopods, in exposures of the Red River Formation. At Pictured Rocks National Lakeshore, in Michigan's Upper Peninsula, you can see rocks full of cephalopods and snails in the Au Train Formation at the tops of Munising and Bridal Veil Falls. Several fascinating caves etched out of Ordovician limestones are scattered across the Upper Midwest. At Niagara Cave, near Harmony, Minnesota, cephalopods, snails, horn corals, and a lone trilobite can be seen embedded in the cave walls, which are made of the Maquoketa Formation. Brachiopods and other invertebrates are exposed in the ceiling of Spook Cave near Giard, Iowa, which is hollowed out of Platteville Formation rocks.

At the Caesar Creek Lake spillway near Waynesville, Ohio, the spillway cuts into the Bull Fork Formation and exposes abundant brachiopods, cephalopods, bryozoans, snails, crinoids, horn corals, and trilobites; some collecting is even allowed (with modest restrictions). In fact, Caesar Creek is only one of several Ohio state parks that allow some fossil collecting in Ordovician Cincinnatian Series formations—the same is true of Stonelick, Hueston Woods, and Cowan Lake state parks. One of the best preserves in all the U.S. for viewing Paleozoic rocks is at Falls of the Ohio State Park near Clarksville, Indiana. Cincinnatian Series rocks there contain innumerable brachiopods, bryozoans, corals, crinoids, clams, snails, trilobites, starfish, and trace fossils. In the western states, public viewing sites for Ordovician fossils are hard to come by, but the Pogonip Group in Death Valley National Park contains abundant large snails.

Trilobite *Eoceraurus trapezoidalis*, Bromide Formation, Oklahoma. Many trilobites evolved defensive spines during the Ordovician (spines have been restored on this specimen). Private collection.

Although its initial designation may have come as the result of a nomenclatural conflict, the Ordovician does indeed represent a distinct and important period in Earth's history. While almost all of our contemporary phyla were present by the end of the Cambrian, the fauna took on a much more modern look during the Ordovician with the extinction of the 'problematica' and the diversification of more advanced types of marine creatures. And the stage was set for the greening of the land, which would proceed apace during the subsequent period, the Silurian.

Chapter Four
The Silurian

The Silurian saw North America largely covered by shallow, tropical seas. This reef in Wisconsin shows eurypterid 'sea scorpions' and nautiloid cephalopods prowling the bottom waters in search of trilobites and other hapless prey. Intelligence in the oceans was increasing, and the predator-prey 'arms race' was in full swing.

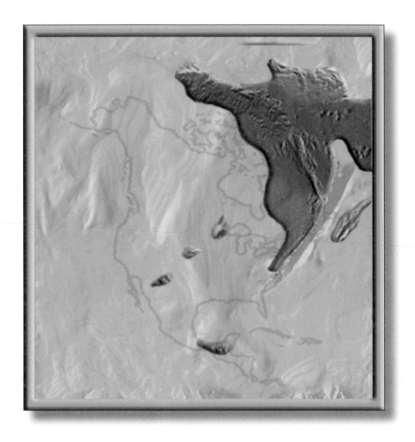

Silurian North America

We come now to the Silurian Period, one of the shortest periods of the Paleozoic Era. One might think that, because of its short duration, the Silurian would also be short on important developments in the history of life. There may be comparatively fewer Silurian rocks and fossils because of the lesser timespan it encompasses, but the Silurian saw the first real terrestrial ecosystems appear—not an unimportant milestone by any means.

Definition and Nomenclature

This third period of the Paleozoic Era was named by Sir Roderick Murchison in 1835. He was working on the stratigraphy of northern Wales, and chose to name this time interval after the Silures, an ancient Celtic tribe that lived along the Welsh-English border in Roman times.

Current dating of the Ordovician-Silurian boundary (start of the Silurian) places this event at 444 million years ago. It was defined in 1984 as the first appearance of the graptolites *Parakidograptus acuminatus* and *Akidograptus ascensus* in an uninterrupted series of black shales exposed at Dob's Linn, Scotland, 1.6 meters above the base of the Birkhill Shale. The Silurian-Devonian boundary (end of the Silurian) is currently dated at 416 mya and was defined in 1972 as the lowest occurrence of the graptolite *Monograptus uniformis* within 'bed 20' at the Klonk section, Barrandian area, southwest of Prague, Czech Republic. The Silurian therefore encompasses a timespan of about 28 million years.

Paleogeography and Paleoclimate

The Ordovician drew to a close with a major episode of glaciation just finishing and Gondwanaland centered on the South Pole. North America occupied tropical latitudes and was rotated about 45° clockwise from its present position, with the equator cutting a swath through northwestern Canada. Variations in the global occurrence of chert formations suggest that the glacial episode lasted about 15 million years, with the last remnants of glaciers melting in the Early Silurian, followed by an interval of sluggish ocean circulation. Seafloor spreading rates were probably low at this time, as continental positions did not change much during the Silurian, and Early Silurian sea levels remained low. A dearth of volcanic rocks worldwide supports the hypothesis of low tectonic activity. Sea levels did rise by the Middle Silurian, however, probably in response to the melting of the last glacial ice caps. This increased the area of epicontinental seas and allowed shallow-water marine life to rebound and refill niches vacated during the terminal Ordovician extinctions.

Some plate tectonic movements did occur slowly during this timeframe, however. The collision of North America with Baltica (the core of modern Europe), which had commenced during the Ordovician, proceeded. The Avalonian continental fragment, bearing the lands that would become the Eastern Seaboard of the United States and parts of Nova Scotia and Newfoundland, slowly drifted north to meet North America. These collisions resulted in the continued rise of the Appalachian and Caledonian mountain chains, whose shed sediments spread into inland troughs to the north and west of the highlands. Siberia also drifted closer to North America, approaching Greenland, Alaska, and the Canadian Arctic isles. Suturing of these continental blocks would later form the supercontinent of Laurasia. The Late Silurian saw the revitalization of volcanic activity and spreading along the midocean ridges.

Trilobites *Gravicalymene celebra*, Springfield Formation, Preble Co., Ohio. Denver Museum of Nature & Science, Denver, Colorado.

Average global temperatures were about 20°C (68°F) during most of the Silurian—significantly higher than today's mean annual global temperature of 15°C (59°F). This value remained fairly constant throughout the Silurian and into the Early Devonian. Thus the Silurian was a time of generally warm, mild climates, although a latitudinal gradient did exist with more temperate climates near the poles. Only the South Pole hosted any land area, however, as most of the northern hemisphere north of the tropics was open ocean.

Significant bands of aridity occurred between the equator and about 40° north and south latitude, leading some restricted continental basins to become huge evaporation ponds, similar to the Persian Gulf of today. One of the largest of these was the Michigan Basin, a huge depression centered on the Lower Peninsula which is filled with salt and alabaster beds from this timeframe. Other Silurian salt deposits are known from New York, Ontario, and Ohio; the 'redbeds' of the Appalachian region formed under terrestrial desert conditions at the same time.

Atmospheric oxygen levels, which were maintained by marine plants—and which had, you may recall, reached half of modern levels by the Ordovician—declined to about one-third the present concentration by the Late Silurian. Atmospheric carbon dioxide, while well above present-day levels by about a factor of three, also declined during the Silurian, possibly as a response to sequestration of carbonate by reef faunas. This CO_2 was likely a factor in maintaining the Silurian 'greenhouse' climate, but O_2 levels would not recover until plants spread across a significant fraction of the continental land surface—something that would not happen until well into the Devonian. Such low atmospheric oxygen caused most ocean waters below about 100 m in depth to be hypoxic or anoxic (low or lacking in oxygen), and thus rocks formed at these depths are sparse in benthic (bottom-dwelling) animal fossils, although many of them contain the remains of planktonic graptolites.

Life in the Silurian

The shallow seas ringing the continents—which covered many parts of North America which are dry land today, such as Kentucky, Ohio, Missouri, Oklahoma, parts of Maryland and Alaska, northern Greenland, and much of the Canadian Arctic Archipelago—hosted vast reefs built up mostly by tabulate corals and stromatoporoid sponges (and to a lesser extent by bryozoans and calcareous algae). The shallows fringing the Michigan and Illinois basins were also the site of extensive reef development; consequently, the Great Lakes region is host to one of the most widespread occurrences of Silurian reefs in the world. Although stromatolites were no longer abundant worldwide, large fossilized stromatolite reefs are known from parts of Alaska and smaller occurrences from New York. These may represent hypersaline environments, as stromatolites today are restricted to lagoons that are too salty for most marine grazing animals. Corals,

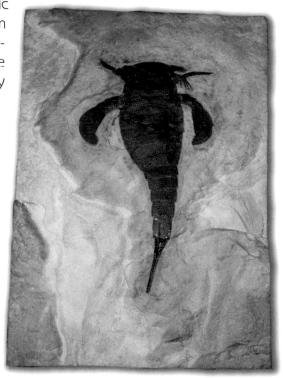

'Sea scorpion' Eurypterus remipes, Fiddler's Green Formation, Herkimer Co, New York. Private collection.

which had first appeared during the Ordovician, became widespread reef-formers at this time. They created huge deposits of limestone in the tropical seas of North America, and these extensive limestones are, in fact, one of the primary characteristics of the Silurian. Coral reefs generally only form in shallow, open-marine environments like the Bahamas today, thus they indicate clear and warm waters.

Although no new major groups of marine invertebrates appeared during the Silurian, there was the continued evolution of existing groups. Brachiopods dominated the reef faunas, sometimes constituting 80% of the fossils we find in those environments; the first spiriferid brachiopods appear at this time. Crinoids underwent a major radiation, and their echinoderm cousins, the blastoids, first appeared. Trilobites had reached an earlier zenith and were now in decline; the agnostid trilobites became extinct. Trilobites did, however, continue to expand the econiches they occupied: some Silurian trace fossils thought to be trilobite walking trails, found in northwest New York State, are superimposed on mud cracks—indicating that those trilobites lived in very shallow water underlain by substrates that were only inundated at high tide. Planktonic graptolites were abundant in the Early Silurian, declining somewhat in importance towards the end of the period. Conodonts remained abundant, and benthic assemblages of bivalves (clams) were widespread. Though nautiloid cephalopods were still common invertebrate predators, they were joined now by a very successful arthropod group—the eurypterids or 'sea scorpions' (which are not really scorpions at all, but look similar). These relatives of the extant 'living fossil' horseshoe crab ranged from a few centimeters long to over 5 feet, and were no doubt formidable predators. Their fossils are commonly found in brackish, back-reef environments, and some of them may even have ventured into fresh waters.

Tabulate coral *Favosites favosus prismaticus*, Michigan. University of Michigan Exhibit Museum of Natural History, Ann Arbor.

Nautiloids began declining in importance during the Silurian, probably due to competition from both eurypterids and an up-and-coming group: the fishes. Jawless fish had become established during the Ordovician, and now underwent a major adaptive radiation. The agnathan (jawless) osteostracans had the anterior region of the body covered in large, bony plates; few fossils of the posterior region of osteostracans are known, probably because this part of the body was covered in a mosaic of scales. The anaspid group, which contains the modern lampreys and hagfishes—the only living agnathans left—became established. However, the major evolutionary innovation which the fishes achieved was the transformation of the anterior gill arch into bony jaws. There is some scrappy evidence that this may have taken place as long ago as the Late Ordovician, but certainly by the Silurian jawed (gnathostome) fishes became well-established, as shown by a wealth of well-preserved body fossils. One very successful early group of jawed fishes was the acanthodians or 'spiny sharks,' which date back to the Early Silurian and are the first fish preserved with teeth and jaws intact. This group sported a streamlined shape, paired fins—protected by sharp, robust anterior spines—and a covering of scales rather than bony plates. Such a morphology suggests that they were built for speed rather than for protection; however, the spines in some acanthodians were large enough that they would have made a nasty meal. The evolution of jaws was to put fish at the forefront of the race for top marine predators, a role they would fill unequivocally during the Devonian. Throughout the Silurian, however, jawed fishes were subordinate in importance to agnathans. Also, Silurian sediments yield the first clear evidence that fish had invaded freshwater environments.

But the most important development of the Silurian was the greening of the land. We have mentioned that Ordovician spores indicate that bryophytes (mosses and liverworts) had already become established. Assuming that their living relatives provide an accurate indication of these nonvascular plants' environmental preferences, they would have been confined to areas of permanent wetness such as the borders of springs, lakes, and streams. Because they lacked a water vascular system, transportation of fluids about the plant was by osmosis, which limited both their habitat and their size. Rare trilete spores are suggestive of an Ordovician origination of vascular plants, but they were uncommon in comparison to the bryophytes. By the Late Silurian, all that had changed. Primitive vascular plants having a 'plumbing system' began to appear all over, most of them assigned to the genus *Cooksonia* and generically called rhyniophytes. These were simple plants lacking roots and leaves, which photosynthesized with their stems and bore their spores in sporangia located on terminal umbels created by repeated bifurcations of the stem. They anchored themselves to the ground with a web of rhizoid filaments, which was less morphologically complex than a true root system. Rhizomatous growth allowed them to spread and form monospecific 'groves.' Although Silurian plants were still small (a few centimeters high) and confined to moist areas, the evolution of a water vascular system set the stage for their descendants to both colonize drier habitats and attain greater heights. It was an evolutionary innovation in the plant kingdom comparable in importance to the evolution of jaws among fishes.

Trilobites *Calymene breviceps*, Waldron Shale, Waldron, Indiana. Note also the small brachiopod at lower right. Private collection.

Many obstacles had to be overcome before plants could move out of the seas and onto the land. First and foremost, they had to be able to resist drying out; they needed a support system to be able to grow upwards against gravity—something that was moot when immersed in seawater; they had to learn to cope with wide daily temperature fluctuations; pores for gas exchange were necessary; they needed a method of obtaining nutrients from the soil base that were not available via osmosis from the air; and they needed to be able to reproduce without constant immersion in water. So it is no wonder that it took millions of years for plants to venture onto the forbidding surfaces of the continents, and millions more for them to colonize drier, upland areas.

Fossils of various species of *Cooksonia* occur throughout the world, with many fine specimens coming from North America—specifically New York and Ontario. The primitive lycophyte (relative of modern clubmosses) *Baragwanathia longifolia* has been found in Silurian sediments in Australia, and is a considerably more advanced type of plant than *Cooksonia*, having true leaves as well as a fully-developed vascular system. So far, such advanced plants have not been uncovered from this timeframe outside Gondwanaland, and may indicate that the hotbed of plant evolution during the Silurian was at temperate, rather than tropical, latitudes. Lycophytes are not known from the northern continents before the Early Devonian.

1990 Canadian fossil stamps. Clockwise from upper left: *Paradoxides*, trilobite, Cambrian; *Eurypterus*, eurypterid, Silurian; *Opabinia*, soft-bodied invertebrate, Cambrian; stromatolite, Precambrian. The fourth stamp is said to be the most widely-circulated image of a Precambrian fossil in the world.

With the colonization of the land by primitive plants came new opportunities for animals. As mentioned in the last chapter, a few fragmentary Ordovician fossils suggest that some arthropods had ventured out of the water back then, but by the Silurian, this was surely so. Centipedes and possible scorpions are known from Late Silurian terrestrial sediments. Body fossils of these arthropods show evidence of sensory and respiratory systems which are fully adapted for life on land, indicating a long prior history that has yet to be uncovered. Late Silurian centipedes and arachnids are predatory forms, implying a simple food web of primary producers (plants), herbivores/detritivores, and predators. Finds of fecal pellets containing fungal hyphae suggest that some of them may also have fed on fungi. The establishment of plants in terrestrial environments provided the basis for the evolution of a more complex food chain—the cornerstone of a true terrestrial ecosystem.

Like plants, however, there were many obstacles for these animals to overcome before they could live on the land. They, too, had to resist desiccation; they needed a respiratory system that could obtain oxygen from the air; again, they needed a support system to replace the buoyancy they had taken for granted in the water; and they needed protection from the sun's ultraviolet rays—as well as something to eat! Plants obliged the fourth of these requirements by exhaling oxygen that was converted to an ozone shield in the stratosphere, and provided the answer to the food problem as well. Arthropods may have been the first terrestrial colonists because their exoskeleton was 'preadapted' (already evolved and then co-opted for another purpose) to provide both support and a water-resistant covering, hurdles which the vertebrates and other invertebrate groups needed to address *de novo*. It would take them until the following period, the Devonian, to do so.

Where Silurian Rocks Are Found

As mentioned previously, there are good exposures of Silurian reefs in northern Greenland and the Canadian Arctic isles, as well as in a circumferential band along the southern shore of Hudson Bay. There are large Silurian outcrops in the province of Manitoba, including the western shore of Lake Winnipeg. There are exposures in the Appalachians from New-foundland to Georgia, but the best Silu-rian rocks on the continent are found in the Great Lakes Region.

Look at the map inside the front cover again. Do you see a "bull's eye" of lavender color centered on the Lower Peninsula of Michigan? These are the Silurian reefs encircling the Michigan Basin, which was a shallow, restricted embayment during that time. These Silurian reefs crop out along the southern shore of the U.P. from Sault Ste. Marie almost to Escanaba; along the western shore of Lake Michigan from Green Bay, Wisconsin to the Illinois-Indiana border; across central Indiana and Ohio as far south as the Ohio River and as far north as the Toledo-Cleveland area on Lake Erie; and in an east-west band across west-central New York,

Trilobites: left - *Calymene niagarensis*; right - *Dalmanites limulurus*. Rochester Shale, Middleport, New York. Private collection.

to Niagara Falls, and turning north across Ontario as far as Southampton and Owen Sound on Lake Huron. The large exposure of Silurian limestones centered on Davenport, Iowa is particularly famous for its articulated crinoids. Small outcrops can be found in the Arbuckle Mountains of Oklahoma.

Unlike rocks of the previous time periods we've discussed, Silurian expo-sures in the Rocky Mountains and the West are few and far between, and they are practically nonexistent south of the Rio Grande.

Where Silurian Fossils Are Found

Some formations and localities produce absolutely spectacular Silurian fossils. Good brachiopods have been reported from the Fossil Hill Formation in Ontario, and the Power Glen Shale is rich in brachiopods, bryozoans, and crinoids. Gastropods (snails) can be found in Nova Scotia's Arisaig Group. The Lockport Formation of Ontario and New York, a resistant dolomite which forms the precipice of Niagara Falls and contains only poorly-preserved stromatolites, corals, and stromatoporoids, protects the underlying Rochester Shale, which yields some of the best-preserved Silurian fossils in North America—including complete cystoids, crinoids, trilobites, and stelleroids (starfish). New York's Reynales Formation is rich in brachiopods, bryozoans, and crinoids, and its Glenmark Shale produces horn corals, brachiopods, and trilobites. Tabulate and rugose corals, clams, snails, and stromatoporoid sponges may be found in the Coeymans and Manlius formations of the Empire State. New York's Pittsford Shale contains many beautiful eurypterids. The Fiddler's Green Dolomite of the Bertie Group, also from New York, in some layers preserves so many eurypterids that it is known locally as the 'sea scorpion graveyard.' This same formation also yields plants, algae, graptolites, ostracods, cephalopods, and phyllocarids (leaf shrimp). Terrestrial Silurian formations in Ontario and New York may sometimes yield specimens of land plants (*Cooksonia*), as does the Phelps Waterlime of eastern New York State. In Pennsylvania, the Keyser Formation is known to contain corals, stromatoporoid sponges, brachiopods, bryozoans, and sometimes cystoids.

'Sea scorpions' *Eurypterus remipes*, Fiddler's Green Formation, Herkimer Co., New York. Denver Museum of Nature & Science, Denver, Colorado.

The Niagara Dolomite of eastern Wisconsin, which is also rich in conodont fossils, is justly famous for its well-preserved trilobites and other Silurian invertebrates. The Niagara Dolomite is the thicker correlate of the abovementioned Lockport Formation, but the fossils here are much better preserved than those on the eastern fringe of the Michigan Basin. The reefs of the Racine Formation, exposed along Lake Michigan in southern Wisconsin and northern Illinois, also contain many invertebrate fossils, as well as abundant graptolites. In Michigan, the Cordell Dolomite of Silurian age produces many wonderful brachiopods, corals, snails, and clams.

Trilobites and trace fossils have been reported from the Poxono Island Formation of eastern Pennsylvania. Tabulate corals, bryozoans, brachiopods, and tiny shelled crustaceans called ostracods can be collected from Silurian outcrops in Pennsylvania's Blair and Union counties. The Brownsport Formation in Tennessee is rich in Silurian invertebrates. If your stomping ground is Indiana, look in the Waldron Shale for some extraordinary crinoids and bryozoans. The Osgood Formation of that state also produces wonderful cystoids in isolated pockets, as well as brachiopods, bryozoans, cephalopods, gastropods, and crinoids. Ohio and Kentucky's Brassfield Formation and Laurel Limestone yield horn corals, brachiopods, bryozoans, crinoids, trilobites, cystoids, and starfish. In some parts of Ohio, the Brassfield is composed almost entirely of crinoid fragments! Oklahoma's Arbuckle Mountains reveal Silurian limestones that produce many spectacular trilobites. Other productive collecting areas in Oklahoma are the western Wichita Uplift and the northeastern Ozark Uplift near Tulsa.

As previously noted, Silurian rocks in the west are sparse, but Idaho hosts several fossiliferous formations. The Roberts Mountain Formation yields many brachiopods, gastropods, and tabulate and rugose corals, whereas the Laketown Dolomite also produces a great variety of corals, brachiopods, and large crinoid stems. The Trail Creek Formation contains abundant graptolites.

Where Can Silurian Fossils Be Seen in the Field?

Because the Silurian spanned so short a time (if time measured in millions of years can be called short!), there are fewer exposures of rocks from this timeframe and fewer places you can visit them as compared with both the flanking periods (the prior Ordovician and the subsequent Devonian). Still, three public preserves deserve mention.

We've met Falls of the Ohio State Park before, and will again, because of the amazing exposures of Paleozoic rocks found there. For the Silurian, these include the Brassfield Formation, Waldron Shale, and Laurel Limestone, which are chock-full of brachiopods, bryozoans, horn corals, trilobites, crinoids, sponges, cephalopods, and trace fossils. This park near Clarksville, Indiana is without question the best place to view *in situ* Silurian fossils in all of North America, and sometimes they even get quarry spoil trucked in and dumped at the rear of the parking lot, to allow visitors to collect some fossils of their own.

Trilobites *Calymene celebra*, Wisconsin. Author's collection.

Death Valley National Park in California has also been previously mentioned as having a good representation of Paleozoic rocks. The Hidden Valley Dolomite there contains abundant Silurian crinoid stems.

A little harder to get to is Drake Island in Glacier Bay National Park, southeast Alaska. But if you do find someone to guide you, or rent a boat, you can see the remnants of an ancient barrier reef that formed when this terrane was an isolated small landmass in between North America and Siberia. The reef was formed by stromatolites and calcareous sponges, and also contains fossils of corals, brachiopods, and snails.

The stratigraphic boundaries of the Silurian are not as clearly marked in North America as they are in Europe and some other areas of the world, generally being gradational rather than abrupt. Whereas the onset of the period is definitively characterized by the divisor of the Late Ordovician glacial episode and extinctions, the Silurian ended quietly from both a geologic and biologic point of view. It was not marked by any catastrophic extinctions, but instead by gradual faunal and floristic change. Nevertheless, the primary evolutionary innovations effected by both plants and animals during the Silurian—the development of a truly terrestrial habit and the appearance of bony jaws, respectively—would be carried farther during the Devonian by at least an order of magnitude, resulting in profound changes in both the marine and terrestrial ecosystems.

Chapter Five
The Devonian

North America occupied tropical latitudes throughout much of the Paleozoic, as this Devonian reef in northern Greenland attests. Often called the "Age of Fishes," an amazing array of shapes and sizes of fish proliferated during this time.

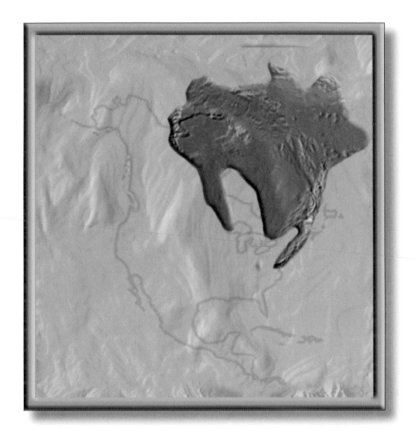

Devonian North America

By now we've covered the early Paleozoic and come to its middle timespan. Plants have ventured onto the land, and with them a few small arthropods. But the vast majority of life has still been confined to the seas, and terrestrial ecosystems are yet rather simple. The Devonian Period would see all that change. Plants ceased to be confined to the margins of watercourses, developed roots and seeds, and spread across the land, forming the first forests; and the earliest tetrapods—four-legged amphibians, evolved from fishes—followed them.

Definition and Nomenclature

The Devonian is the fourth period in the Paleozoic Era, and occupies the middle position—and timespan—of the "Era of Ancient Life." The initial work on these rocks was done in the 1830s by Henry Thomas de la Beche, in Devonshire, England, from whence the name of the period is derived.

The beginning of the Devonian (Silurian-Devonian boundary) was defined in 1972 as the first appearance of the graptolite *Monograptus uniformis* in a series of rhythmically alternating limestone beds and calcareous shales found southwest of Prague, Czech Republic—specifically 'bed 20' in the Klonk Section, Barrandian Area. Current dating places this boundary at 416 mya. The Devonian-Mississippian boundary (end of the Devonian) was chosen in 1990 as the first occurrence of the conodont *Siphonodella sulcata* at the base of 'bed 89' at La Serre, Montagne Noir, Cabrières, in southern France. The geologic unit in which this occurs is called the "Klippen of Cabrières" and is a sequence of predominantly biodetrital oolitic limestones within a matrix of marine shales, presently dated at 359 mya. Thus the Devonian Period encompasses a timespan of about 57 million years.

Paleogeography and Paleoclimate

You will recall from our discussion of Silurian paleogeography that North America was situated in tropical latitudes with the equator cutting through north-western Canada, and the continent rotated about 45° clockwise from its present position. Baltica—the core of modern Europe—had sutured itself to North America's Labrador-Greenland coast, raising the Caledonian mountain chain. Together North America and Europe are frequently called Laurasia or the "Old Red Continent," after Britain's Old Red Sandstone of Devonian age and similar ruddy-colored formations that formed continent-wide at the same time. The reddish color of these rocks reflects the absence of mature soil development and effects of the generally arid climate that prevailed across the emergent lands of Laurasia throughout much of the Early and Middle Devonian.

North America's geographical position didn't change much during the Devonian, but elsewhere continental drift was both adding to our landmass and beginning the chain of events that would lead to the formation of the supercontinent of Pangea in Late Paleozoic times. The Appalachians continued to rise in what is called the Acadian Orogeny, as the continental fragment of Avalonia secured its position along the New England coast. Subduction zones flanked both the present east and west coasts of North America, and a chain of exotic terranes, the Antler Arc, approached the western continental shelf and welded itself on, beginning the series of accretionary mountain-building events that has since produced much of North America west of the Rocky Mountain Front. The Antler Arc added lands along the Pacific coast of the continent from British Columbia to southern California. Siberia still lay closeby north of the Canadian Arctic Archipelago, and Gondwanaland—an agglomeration of all the southern continents—continued to occupy high southern latitudes, although it underwent some counterclockwise rotation about a center in what is now Australia.

Throughout the early and mid-Devonian, climates worldwide were warm and sea levels high. There was extensive aridity across much of North America and salt and gypsum deposits formed in restricted lagoons and in sabkhas (evaporite

Fish *Bothriolepis canadensis*, Escuminac Formation, Québec. Denver Museum of Nature & Science, Denver, Colorado.

basins above the high-tide line). Much of our continent was underwater, and these warm, shallow epicontinental seas were ideal for the development of extensive coral reefs.

In the Late Devonian, a mass extinction occurred (the so-called Frasnian-Famennian event). Shallow marine life in tropical latitudes was decimated, but high-latitude and terrestrial life was less affected. The F-F event, which began around 375 mya (16 million years prior to the end of the Devonian), ranks among the "big five" mass extinctions of all time. Reefs were devastated: tabulate and rugose corals were hit hard, and stromatoporoids went extinct. Reef building was so strongly affected that it did not fully recover until the appearance of modern hexacorals during the Triassic. Many fish groups—especially jawless and armored ones—also declined in abundance. Trilobites, already in trouble, were almost wiped out. In all, 70% of the taxa of marine invertebrates did not survive past the end of the Devonian.

Many hypotheses have been offered to explain these extinctions, from asteroid impacts to marine anoxia to major volcanism. The most convincing evidence, however, supports the theory that ultimately, plants were to blame. They removed so much CO_2 from the atmosphere that a major global cooling was initiated. Computer simulations, as well as analyses of ancient soils, confirm this. How did they manage it? Well, for one thing, the greening of the land led to the burial of organic carbon in terrestrial sediments. But also, the actions of roots had a major effect on the global carbon cycle: by secreting into the groundwater some of the carbon dioxide the plant had absorbed from the air, roots produced carbonic acid in the soil. This combined with magnesium and calcium derived from rocks to form insoluble carbonates. These plants' actions resulted in a reduction in atmospheric CO_2 sufficient to bring about a drop in equatorial ocean surface temperatures from 32°C (90°F) to 25°C (77°F). This would explain why tropical and shallow water faunas were most severely affected, and why high-latitude organisms migrated towards the equator at this time. The cooling was enough to set off another round of glaciation on Gondwanaland, as evidenced by contemporaneous glacial deposits in Brazil. Additionally, the exhalation of oxygen by the first trees, combined with this sequestration of carbon dioxide, allowed for conditions conducive to the first forest fires, evidence of which is preserved in the fossil record as thin charcoal horizons.

Life in the Devonian

Because, as we have noted, the Silurian ended quietly without any major biological catastrophes, the dawn of the Devonian was likewise

Trilobite *Phacops rana crassituberculata*, Silica Shale, Silica, Ohio. Private collection.

Trilobite *Huntonia lingulifer*, Haragan Formation, Arbuckle Mountains, Oklahoma. Private collection.

marked by only modest change. Marine reefs reached a peak in their development, though Devonian reefs differed little from Silurian ones in being formed by tabulate and rugose corals*, stromatoporoid sponges, and calcareous algae.   Brachiopods, especially spiriferids, were abundant, as were starfish, crinoids and blastoids, and bivalve and cephalopod molluscs (the first ammonoids—cephalopods with coiled shells and complex sutures—appear at this time).   Also common were eurypterids (the 'sea scorpions' we met in the last chapter), bryozoans, and conodont animals; the first siliceous sponges also made their debut. Graptolites and trilobites were on the wane. Some of the decrease in trilobite diversity may have been due to increasing predation from cephalopods, eurypterids, and the fast-evolving fishes.

Brachiopod *Spirifer macronatus*, Ludlowville Formation, Smoke Creek, New York. Private collection.

The great success story of the Devonian seas was the fishes. In fact, the Devonian is often called the "Age of Fish" precisely because they radiated so extensively at this time. Jawless fish remained common, but were surpassed in diversity and abundance by the many new types of fish with jaws. Acanthodians (the so-called 'spiny sharks'—although they weren't sharks at all) reached their peak, starting out as marine creatures and later becoming exclusively denizens of fresh waters.   The first true sharks appeared on the scene around 400 mya. Shark-like scales (tesserae) are known from as far back as the Ordovician, but no teeth are associated with them, possibly indicating that shark ancestors were toothless and maybe even jawless. The first shark teeth appear in the Early Devonian of Antarctica, perhaps indicating that the waters around Gondwanaland were the birthplace of this very successful group. Sharks achieved a cosmopolitan distribution by the end of the Devonian, with especially good remains coming from black shales in Ohio and Pennsylvania.   The placoderms—an armored, predatory group that included the giant arthrodire *Dunkleosteus* from the Cleveland Shale of Ohio and New York—dominated the Devonian seas, but became extinct by the end of the period.

The real action in the fish world (from the perspective of the terrestrial vertebrates who are reading this), though, was in the evolution of the osteichthyans or modern bony fish. This group contains the actinopterygians or ray-finned fish—which constitute most of the fish in today's waters—and two other groups: the Dipnoi or

Trilobite *Kettneraspis williamsi*, Haragan Formation, Arbuckle Mountains, Oklahoma. Private collection.

---

* Just as an interesting aside, Devonian horn corals have actually been used as records of ancient astronomy. Like modern corals, extinct corals added a fine layer of calcium carbonate to their calyces (living chambers) every day. These bands also cluster in annual increments, so by counting the fine layers, the number of days in a Devonian year can be calculated. (The Earth's rotation has been slowing down over time due to the friction of the tides.) It turns out to be 398 days of 22 hours each—exactly the same as predicted by astrophysicists!

lungfishes (a group with three species remaining today), and the Crossopterygii or lobe-finned fish (represented now by the two living species of the coelacanth *Latimeria*). The oldest osteichthyan fossils are 410 million years old, and the first articulated remains a mere 10 million years younger than that. The oldest lobe-fins come from Prince of Wales Island in Arctic Canada. Osteichthyans are very well known from Devonian rocks, and fossils indicate that ray-finned fish first evolved in fresh waters, only later spreading into the sea. An important osteichthyan innovation was the evolution of a swim bladder, originally used for achieving hydrostatic equilibrium, but which later could be modified for use as a lung. We will return to this facet of our discussion a bit later, after we've looked at what was happening in the plants' conquest of the land.

The Late Silurian and Early Devonian landscape was still bare in areas lacking abundant water. But on mudflats and near the banks of permanent water sources, simple nonvascular plants similar to mosses and liverworts had already gotten a good foothold. The rhyniophytes were still widespread, frequently forming monotypic stands suggestive of spread by creeping rhizomes. Although we've mentioned that early vascular plants (tracheophytes) are known from Gondwanaland by the Late Silurian, it isn't until the Early Devonian that they are found on the northern continents. Well-preserved specimens of the early lycophyte *Baragwanathia* appear in the Sextant Formation of Ontario. Other groups of vascular plants that made their debut in the Early Devonian include the zosterophylls and the trimerophytes. In general habit the zosterophylls were similar to the rhyniophytes, being small plants anchored by rhizoids and spreading via rhizomes. The trimerophytes were taller, with a main axis that branched repeatedly, but still lacking a real root system. Maine's Trout Valley Formation preserves extensive 'groves' of *Pertica*, a plant belonging to the trimerophyte group. Most of these plants were small, not exceeding a meter in height, and many were only a few centimeters tall.

Armored fish *Dunkleosteus terrelli* (cast), Cleveland Shale, Ohio. Denver Museum of Nature & Science, Denver, Colorado.

Much bigger conquests were to follow. As well as the development of an internal water-transport system, the tracheophytes evolved the first true roots. Roots increased the efficiency with which plants could anchor themselves to the substrate as well as absorb water and nutrients from it. The actions of roots also led to the appearance of the first well-developed soils. Plants were now in a position to colonize areas further away from permanent water, and they wasted no time in doing so. Lycopods (relatives of today's clubmosses or 'ground pine') became very common by the mid-Devonian, and the develop-

ment of structurally supportive stems allowed them to greatly increase in size, attaining heights of a couple of meters.

Fish *Eusthenopteron foordi*, Escuminac Formation, Québec. Denver Museum of Nature & Science, Denver, Colorado.

The first plants large enough to be called trees appeared by the mid-Devonian, around 385 mya. Petrified stumps uncovered in the 1870s in riverbed exposures of the Oneonta Formation near Gilboa, New York have long been recognized as an example of one of these early forests, but no one knew what the trees really looked like until just recently. In 2007 it was announced that fossils found closeby showed the foliage attached to wood matching the Gilboa material. Known as *Eospermatopteris*, these trees reached some 8 meters (26 feet) in height, and were topped by a crown of repeatedly bifurcating (forking) branches that somewhat resembled a bottlebrush—but they still had no true leaves. Forests of *Eospermatopteris*, therefore, were probably more 'open' and less shady than those we are accustomed to today.

By the Late Devonian, forests had appeared in the equatorial regions of what is now Arctic Canada. These contained a variety of plants which more or less simultaneously developed an arborescent habit: ferns, lycopsids, and progymnosperms, the latter sporting the first *bona fide* wood. The first sphenopsids (horsetails) are found in Late Devonian rocks in Alaska. By this time, plants with true leaves appeared; this was a big advantage, as it made them much more efficient at capturing sunlight than their predecessors had been, with only green, forking stems. A particularly successful genus of progymnosperm tree was *Archaeopteris*, which often grew in monotypic stands and reached heights of 10-30 m (33-98 ft), with a trunk diameter of up to 1.5 m (5 ft) at the base. *Archaeopteris* formed gallery forests and dense patches on coastal and riparian floodplains. It had pinnate fronds superficially similar to those of a fern and was heterosporous (having two sizes of spores), the first step towards the development of seeds. It has been suggested that this tree may have been deciduous, shedding its fronds during the dry season, which would account for the abundance of its fossils. *Archaeopteris* may have been an early precursor of the pteridosperms, or 'seed ferns,' which were to become so common in the 'coal swamps' of the succeeding periods. The first pteridosperms, which were gymnosperms having true seeds, appeared near the close of the Devonian. The development of seeds freed these plants from the need for a water source for the germination of spores, and allowed them to colonize drier, upland habitats.

By this time, plants had diversified into a number of ecotypes, including shrubs, trees, vines, ground cover, and swamp-tolerant varieties. With the spread of plants over much of the landscape and the development of stratified forests, terrestrial ecosystems were fast evolving some complexity. Fossils of *Psilophyton* from the Gaspé Peninsula, Québec, show damage suggestive of feeding by insects with sucking mouthparts, although clear evidence of actual herbivory does not appear until the Pennsylvanian—early on, most plant material probably entered the food chain via detritivory (animals feeding on shed leaves and other decomposing plant matter). The main food chain seems to have consisted of plants-detritivores-carnivores; it has been suggested that terrestrial arthropods had not yet evolved the ability to digest the lignin in plant tissues, although this is difficult, if not impossible, to prove.

Horn coral *Zaphrentis prolifica*, Hungry Hollow Formation, Arkona, Ontario. Author's collection.

The Silurian had seen forays onto land by myriapods (centipedes and millipedes) and arachnids (scorpions, spiders, and mites); by the Devonian, wingless insects joined them, and some eurypterids may have spent part of their time out of the water. Recent finds of 400 million-year-old mouthparts of insects which exhibit features found predominantly in winged forms, such as ribbed structures and a double hinge (which increases chewing power), suggest that flying insects may also have been present during the Devonian, although fossils of the wings themselves have not yet been found.

Towards the end of the Devonian, a truly momentous change occurred in the land fauna: the appearance of the first amphibians. Although it is still debated whether these evolved from the lungfishes or the lobe-fins, one or the other of these groups surely contains the ancestor of all tetrapods (four-legged vertebrates). I favor the view that amphibians evolved from osteolepiform lobe-finned fish, based on the fact that these fish possessed a labyrinthine infolding of the tooth enamel indistinguishable from that of early amphibians, as well as the relatively completely-known skeletal series which seems to link the 365 million-year-old lobe-finned fish *Eusthenopteron* from the Escuminac Formation of Québec with the early amphibians *Ichthyostega* and *Acanthostega* found in Late Devonian sediments of East Greenland. Nevertheless, the matter remains a contentious one among experts working in the field.

Tree *Archaeopteris macilenta*, Pocono Sandstone, Pennsylvania. Denver Museum of Nature & Science, Denver, Colorado.

Like arthropods before them, vertebrates faced a host of challenges in moving from a watery

Trilobite *Huntonia oklahomae*, Haragan Formation, Oklahoma. Private collection.

environment onto the land. Remember how we saw that the bony fish evolved a swim bladder that was ready-made for conversion into a lung? This put them a step ahead of the game, before they even made their initial forays onto the land. In fact, many of the adaptations needed for terrestrial life seem to have evolved in these fishes before they actually made the transition to an amphibious lifestyle. In contrast to the old school of thought, which had fish wiggling across the land to escape from pools of water that were drying up, current thinking is that amphibians evolved from fish that only occasionally hoisted themselves onto the shore before returning to their home in the water. (In the Florida Everglades today, fish and other aquatic animals in drying pools stay there with the little water that remains.) The lobe-finned fish already possessed skeletal elements supporting their fleshy fins which can be identified as a humerus, radius, and ulna; these needed merely to be strengthened and modified to become the bones of a tetrapod limb. In fact, the living coelacanth can move its pectoral and pelvic fins independently in a manner similar to tetrapod locomotion (this has been filmed in its deep-sea habitat). Distal elements became fingers and toes, although early amphibians had seven or eight of them. The skeleton of *Ichthyostega*, one of the best-known early amphibians, shows well-developed limbs attached to an otherwise very fishy body. It still possessed a notochordal brain-case, subopercular bone, and ray-supported tail fin—all fishy features. Lobe-fins and lungfishes declined at the end of the Devonian, possibly due to competition from these newly-evolved amphibians. Early tetrapods probably spent most of their time in the water, only occasionally venturing onto the land (perhaps to bask in the sun as an aid to digestion); sedimentary evidence indicates that they lived mainly in vegetation-choked rivers. The gallery forests that lined their riparian habitats would have provided sufficiently moist environments to allow these forays to be relatively comfortable. Like all amphibians, they were still entirely dependent upon water for reproduction; nonetheless, the emergence of vertebrates from the water would have profound consequences for the future of life on our planet.

In addition to their skeletal fossils, we have evidence that tetrapods really did venture onto the land in the form of footprints. Although these are scarce and sometimes ambiguous, salamander-like tracks have been identified in Upper Devonian rocks in Pennsylvania.

Where Devonian Rocks Are Found

Devonian rocks are exposed at the surface in many areas of North America, particularly east of the Mississippi River and in the Arctic regions. There are outcrops along portions of the southern shores of Hudson and James bays, in parts of Alaska including the panhandle and the North Slope, in the Yukon, the western Northwest Territories, the Canadian Arctic Archipelago, and eastern Greenland. Much of southwestern Ontario, between Lakes Erie and Huron, is underlain by Devonian bedrock. Outcrops can be found across most of the southern half of New York state and parts of northern Pennsylvania all the way west to the shore of Lake Erie. Others are exposed in eastern and western Maine, northern Vermont, up the Gaspé Peninsula in Québec, and into Nova Scotia. Portions of Massachusetts host Devonian bedrock. Some rocks of this age also occur along the west side of the Appalachians, particularly in Virginia, West Virginia, Tennessee, and northwest Georgia.

There is also a large swath of Devonian sediments north-to-south through the middle of Ohio, and in an east-west band across northwestern Ohio through Indiana as far as the Illinois border. Outcrops can be found in an arc across the northern portion of Michigan's Lower Peninsula from Manistee to Alpena, and in a north-northwest-trending band across eastern Iowa roughly from Cedar Rapids to Mason City. Some Devonian outcrops occur in the western states, but they are few and far between. Further from home, and south of the border, Devonian bedrock is exposed in small outcrops in northwestern Mexico (Sonora) and in a generally east-west band from the Mexican state of Chiapas across Guatemala and into Honduras.

Where Devonian Fossils Are Found

Many famous fossil-producing localities are contained within Devonian sediments. Good acanthodian fish fossils come from the DeLorme Formation of Canada's Northwest Territories and the Late Devonian Escuminac Formation of Québec, which also yields coelacanth fossils. The Fram and Nordstrom Point Formations of Ellesmere Island, Canada, produce a fish fauna that includes agnathans, placoderms, and lobe-fins. The Catskill Formation at Red Hill, Pennsylvania produces some of the earliest-known tetrapods, as well as fossil lobe-finned, placoderm, and acanthodian fishes, plants, and invertebrates. The Chadkoin Formation, also in Pennsylvania, contains some fish fossils from both arthrodires and acanthodians. The Cleveland Shale of Ohio, northwestern Pennsylvania, and New York is renowned for its fossil fishes, including placoderms and sharks, and is known worldwide for its specimens of the giant 'terror fish' *Dunkleosteus*.

Crinoids *Arthroacantha carpenteri*, Silica Formation, Sylvania, Ohio. Private collection.

The Arkona Shale and Hungry Hollow Formation of southern Ontario yield beautiful crinoids, horn corals, brachiopods, and the occasional trilobite or starfish, and the Bois Blanc Formation of Ontario and Michigan is rich in brachiopods and corals. Michigan's state stone, the 'Petoskey stone,' is a tabulate coral commonly found in rivers and on beaches in the northern Lower Peninsula; fragments of the underlying Devonian reefs were ground off and smoothed by glacial action. Rounded river pebbles of these 'Petoskey stones' are particularly attractive when polished, and are commonly seen in souvenir shops. Michigan and Ohio's Silica Shale is renowned for its brachiopods and trilobites, and occasionally also yields phyllocarids (leaf shrimp), corals, and echinoderms; Ohio's Columbus Limestone produces nice gastropods and bryozoans. The Jeffersonville Limestone straddling the Indiana-Kentucky border is chock-full of brachiopods, stromatoporoids, corals, bryozoans, blastoids, crinoids, trilobites, clams, and snails. In Tennessee's Ross Formation, brachiopods are extremely common. Maryland's Needmore Shale contains some nice trilobites. New York's Ludlowville and Moscow formations frequently yield an abundance of trilobites, nautiloids, brachiopods, crinoids, corals, and bryozoans. Although the state of Maine is not renowned for its fossils, some brachiopods and clams can be found in the Chapman Sandstone there.

Petoskey stone *Hexagonaria percarinata,* river pebble, northern Michigan. Author's collection.

The Trout Valley Formation of Maine contains good fossils of early land plants, as do the Sextant Formation of Ontario and the Torbrook and Murphy Brook formations of Nova Scotia. Tennesee's Chattanooga Shale sometimes produces silicified logs, called *Callixylon,* which are the trunks of *Archaeopteris* trees. New York's Oneonta Formation contains an entire 'petrified forest' of *Archaeopteris* and other progymnosperms, including *in situ* stumps.

Oklahoma's Haragan Formation yields some of the most spectacular trilobites known from all of North America, beautifully preserved in three dimensions and still sporting long, delicate spines. Small Devonian outcrops in the West can also be highly fossiliferous: the Beartooth Butte Formation of Wyoming contains carbonized remains of both rhyniophyte and lycopsid plants, fragmentary jawless fish, scorpions, and eurypterids; the Chaffee Formation, exposed in Glenwood Canyon, Colorado, yields well-preserved brachiopods, as does the Mount Hawk Formation of Alberta's Rocky Mountains. Many well-preserved crinoids, bryozoans, clams, snails, corals, and large fish bones have been found in Idaho's Three Forks Formation.

Where Can Devonian Fossils Be Seen in the Field?

Public preserves where Devonian fossils are exposed are much more common than they are for most other periods of the Paleozoic. Most of these parks showcase marine invertebrates, but there are two in particular where you can see fish and plants.

The first of these is Miguasha National Park in Québec. Situated on the south shore of the Gaspé Peninsula, near the town of Carleton, this park is another World Heritage Site, though not as well-known as the Cambrian Burgess Shale. Exposures of the Escuminac Formation in the cliff facing Baie de Chaleurs are home to twenty-some species of fish and early amphibians, including the famous *Eusthenopteron*. The rocks also contain many invertebrates and some plant fossils.

Gastropod *Platyceras multispinosum*, Columbus Limestone, Ohio. Private collection.

Moving south, Baxter State Park in northern Maine hosts exposures of the Trout Valley Formation, which have produced countless specimens of the trimerophyte plant *Pertica quadrifaria* (which is also, by the way, the state fossil). Other fragmentary plant material can also be found there. This park is a real treat for fossil aficionados in Maine, who are accustomed to mostly fossil-free granite in their own back yards.

The Penn-Dixie Paleontological and Outdoor Education Center near Hamburg, New York is situated on outcrops of the Windom and Wanakah shales, both Devonian marine units. Fossils found there are brachiopods, bryozoans, clams, snails, crinoids, cephalopods, and trilobites. They even have a program by which you can dig for, and keep, your own trilobites.

Pennsylvania's Swatara State Park, near Lebanon, preserves rocks of the Mahantango Formation from which come brachiopods, bryozoans, and a few trilobites and crinoids. Again, here limited collecting is allowed.

The Rock Glen Conservation Area just north of the miniscule berg of Arkona, Ontario is well worth a visit. Dense hardwood forests shade the banks of the Ausable River, which cuts a gorge through the Widder Beds, Hungry Hollow Formation, and Arkona Shale, all Devonian marine units. Abundant brachiopods and horn corals, and the occasional trilobite, wash down from the cliffs into the creek, completely free of matrix. You are allowed to collect one specimen of each different species you find, but are asked to leave the rest for other visitors.

Brachiopods from Rock Glen Conservation Area: upper - unidentified; lower - *Mucrospirifer arkonensis*. Hungry Hollow Formation, Arkona, Ontario. Author's collection.

Fossil Preserve Park at Sylvania, Ohio is a little different. The fossils here are not *in situ* (in their natural place), but loads of the Silica Shale are trucked in from local quarries and dumped there, specifically to allow visitors to collect their own fossils without the dangers and liability of having access to the working quarries. Not the most scenic of localities, perhaps, but an interesting way to solve the problem of the public's gaining access to this world-famous rock unit. Trilobites, brachiopods, corals, and crinoids are all found in the Silica Shale. At present, only hand collecting (without tools) is allowed—no hammers, chisels or the like.

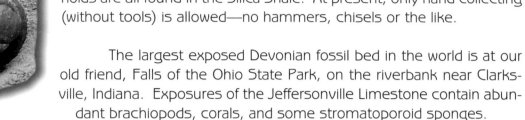

Trilobite *Dechenella lucasensis*, Silica Shale, Silica, Ohio. Private collection.

The largest exposed Devonian fossil bed in the world is at our old friend, Falls of the Ohio State Park, on the riverbank near Clarksville, Indiana. Exposures of the Jeffersonville Limestone contain abundant brachiopods, corals, and some stromatoporoid sponges.

And last but not least, the Devonian Fossil Gorge along the Coralville Lake spillway at Iowa City, Iowa showcases wonderful examples of brachiopods, crinoids, bryozoans, corals, and trilobites of the Coralville Formation. The rocks were exposed during a 1993 flood of the reservoir. Here, you can actually walk along a Devonian sea floor!

The Devonian saw great changes, both on the continents and in the seas. Never again would the land be an alien and forbidding place—and our distant ancestors were there to explore it.

Chapter Six
The Mississippian

The Mississippian was a time of warm, shallow seas and extensive limestone deposition across North America. Fish like sharks and coelacanths cruised over forests of crinoids or 'sea lilies,' like this one in Missouri, which sheltered communities of brachiopods, ammonoids, and the few remaining trilobites.

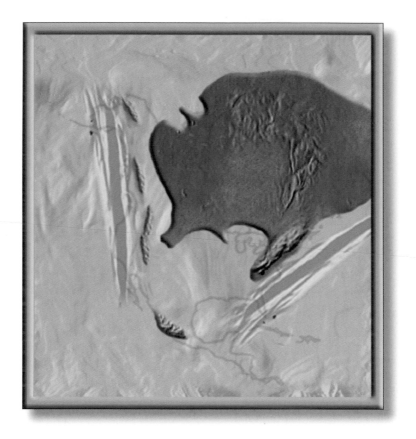

Mississippian North America

We are now a bit more than halfway through the Paleozoic Era, and have seen tremendous changes in the floral and faunal inhabitants of our native continent and its peripheral shelves. Although, by this time, the land surface was no longer barren, the times up till now have seen much of North America flooded by shallow seas. The Mississippian was the last time during which vast areas of our continent would be inundated. Most Mississippian rocks in North America are marine limestones. As we move forward, though, our story will increasingly be one of terrestrial animals and plants.

Definition and Nomenclature

In most of the world, the Mississippian (and the subsequent Pennsylvanian) are not afforded Period status. They are considered Epochs—the next-lowest division—by the International Commission on Stratigraphy (ICS), corresponding closely with the boundaries of the Lower and Upper Carboniferous Period, respectively. The reason for this is that in Europe and most of the rest of the world, there is no distinctive difference in the rocks and environments between the Mississippian and the Pennsylvanian, so workers there use the older, English term Carboniferous (meaning 'coal-bearing') for the entire timeframe. In North America, by contrast, the Mississippian and Pennsylvanian differ markedly from each other both lithologically and in their fauna and flora. As aforementioned, during Mississippian time, much of North America was covered by calm, shallow seas which left extensive deposits of limestone; instead, terrestrial 'coal swamps' dominate Pennsylvanian rocks. This book is about the paleontology of North America, and geologists here afford both the Mississippian and the Pennsylvanian full Period status—thus we shall do so as well, and devote a separate chapter to each.

The name Mississippian was proposed by American geologist Alexander Winchell in 1869, as a reference to the Mississippi River valley, where marine limestones are abundantly exposed. In 1891, the U.S. Geological Survey formally divided the Carboniferous into the Mississippian and Pennsylvanian; it was proposed in 1906 that these be raised to systemic (Period) rank, a practice which has been followed by North American workers ever since. The Mississippian and Pennsylvanian names (as Epochs) were ratified by the ICS in 2000.

The Devonian-Mississippian boundary (beginning of the Mississippian) was defined in 1990 as the lowest occurrence of the conodont *Siphonodella sulcata* at the base of 'bed 89' at La Serre, Montagne Noir, Cabrières, southern France. Current dating places this boundary at 359 mya. The Mississippian-Pennsylvanian boundary (end of the Mississippian) was defined in 1996 at the lowest occurrence of the conodont *Declinognathodus nodiliferus (sensu lato)* 82.9 m above the top of the Battleship Wash Formation in Arrow Canyon, southern Nevada. This is currently dated at 318 mya, such that the Mississippian encompasses a timespan of about 41 million years.

Mississippian crinoids, Crawfordsville, Montgomery Co., Alabama. New Mexico Museum of Natural History, Albuquerque.

Paleogeography and Paleoclimate

We left the Devonian with North America occupying tropical latitudes and rotated about 45° clockwise from its present orientation. The Mississippian saw our continent drift somewhat northward, so that the equator became positioned along a line from the southern tip of Nevada through James Bay in Canada, cutting the continent approximately in half. North America remained firmly sutured to Baltica (Europe) in the northeast, and Siberia drifted northward and eastward to a position on the far side of Baltica, with which it would later collide to form the Ural Mountains. Gondwanaland (an agglomeration of all the southern continents) drifted northward until Australia occupied southern subtropical latitudes, but with portions of Africa and South America still firmly straddling the south pole. As Gondwana approached Laurasia (North America plus Baltica), the Rheic Ocean began to close, and the first hints of the shape of another supercontinent-to-come, Pangea, emerged. Throughout the Mississippian, large areas of North America continued to be flooded by warm, shallow seas, the last time that a majority of the continent was submerged beneath the waves. In the west—near the California-Nevada border—the Antler Arc had met the continental shelf, raising the Antler Mountains flanked on the east by a foreland basin. Much of the Western Interior was inundated, as were large expanses of the midwestern and southern states. The first hints of what was later to become Mexico appeared as a few isolated islands. The Appalachians provided some relief in the east, and the Canadian Shield region remained above water. The end of the Mississippian saw extensive mountain-building (associated with the early stages of the formation of Pangea) in New Brunswick, Nova Scotia, the southern Appalachian region, and the southwestern United States, as well as in Europe.

Shark spine Ctenacanthus *sp., Iowa. Denver Museum of Nature & Science, Denver, Colorado.*

Because of the waxing and waning of glaciers on Gondwanaland, sea levels, while generally high, fluctuated, and a latitudinal gradient existed with cooler climates in the high southern latitudes. Siberia and Kazakhstania occupied temperate latitudes in the northern hemisphere, but other than that, northern lands were tropical; a vast ocean, the Panthalassic (forerunner of the modern Pacific), covered most of the northern hemisphere and half of the southern. The tempering effects of this huge body of water kept most continental climates fairly equable, save for the inland areas of Gondwana—continental climates in general were warmer, more humid, and less seasonal in the Mississippian than they are today. Lack of growth rings in Mississippian tree trunks attests to the generally less seasonal climate of the time. Subduction zones continued to flank the eastern and western coasts of North America; as the Rheic Ocean closed, there was renewed uplift in the Appalachian region, which was the only part of the eastern United States to remain above water at this time. The presence of abundant crinoids in Mississippian limestones indicates that the epicontinental seas were warm and clear—crinoids are filter-feeders which cannot tolerate turbid water conditions, and which require the high concentrations of dissolved calcium carbonate found only in warm waters to build their skeletons. The Mississippian is sometimes called the "Age of Crinoids," because some limestones are composed primarily of the disarticulated columnals of these stalked echinoderms. Such rocks are called encrinoidal limestones. It has been calculated that the number of crinoid animals necessary to account for the abundance of their columnal plates in one formation in Kentucky alone is 28×10^{16} (280 quadrillion)! Other limestones are composed primarily of oolites, which are small spheres of calcite precipitated out of tropical waters and shaped by wave action. Modern oolites are still forming in the Bahamas today.

Shrimp, Bear Gulch Limestone, Montana. St. Louis Science Center, St. Louis, Missouri.

Trilobite tracks, Glen Dean
Formation, Perry Co.,
Indiana. Author's collection.

The Mississippian climate of the emergent portions of Laurasia was warm and sometimes seasonally dry, with a trend towards cooler and wetter conditions as the Pennsylvanian approached. Concomitant with this was a general drop in sea level leading to more emergent lands. Thin, alternating beds of limestone and shale indicate that the seas were becoming muddier as the lands emerged. Swamp coals began to form in the latest Mississippian during the wettest stages; these are recorded in the stratigraphy of some areas of eastern North America as thin, intercalated beds of coal and shale. But we will have to wait until our look at the Pennsylvanian for an in-depth discussion of these coal swamps.

Life in the Mississippian

Reef-building organisms had been decimated in the Late Devonian extinction, and there was a dearth of framework-generating species throughout the Mississippian. Nevertheless, the seafloor was abloom with these unimaginable numbers of crinoids, which must have formed great 'forests' on the bottom. Brachiopods and blastoids were extremely numerous also. Large foraminifera called fusulinids first appeared near the end of the Mississippian. Trilobites were on their way out and are only represented by a few species throughout the remainder of the Paleozoic. Among molluscs, ammonoids rebounded, and some nautiloids were still around, as were clams and snails. The first ceratitic ammonoids, bearing more complex suture patterns than their goniatitic cousins, appeared. Armored fish were mostly replaced by faster-swimming varieties covered only in scales—testimony to the value of speed in escaping the increasingly more efficient predators (including sharks). The Bear Gulch Beds in Montana yield one of the most diverse fossil fish faunas in the world, many of which are articulated and even preserve soft parts. This plattenkalk (fine-grained limestone) represents a shallow basin or estuarine environment and has given us a unique glimpse into a Mississippian ecosystem replete with acanthodians, sharks, coelacanths, skates, and ray-finned fish, along with ammonoids, conodonts, bryozoans, gastropods, brachiopods, sponges, shrimp, worms, nautiloids, and starfish. What a cast!

Photosynthesis increased sharply with the spread of terrestrial vegetation in the Devonian, and continued to do so as complex floras occupied more and more of the land surface in the Mississippian. You'll remember from the last chapter that plants were fast sucking the carbon dioxide out of the atmosphere. The effects of this are seen in the number of stomata (respiratory pores) on the leaves and stems of Mississippian plants vs. modern plants. Stomatal density correlates inversely with CO_2 concentration in the air: the less CO_2 available, the more stomata are necessary for the plant to supply its respiratory needs. Carbon dioxide decreased from the Late Devonian through the end of the Pennsylvanian, reaching a level comparable to today's 383 ppm around the Pennsylvanian-Permian boundary. Conversely, oxygen concentrations were on the rise, allowing for frequent forest fires and possibly accounting for the gigantism seen in some insects and amphibians in the Late Mississippian and Pennsylvanian. These creatures have less-efficient respiratory systems than higher animals and can grow to larger sizes if more oxygen is available.

The large forests of *Archaeopteris* trees were extinguished at the end of the Devonian, and it took a while before any other types of trees of comparable size evolved. In the Early Mississippian, most lycopods and ferns were 2 meters tall at most, and much of the land surface was covered by what would today be called 'weeds' (although of completely different types from modern herbaceous angiosperms, which wouldn't make an appearance until the Cretaceous). As the period progressed, lycopod trees became taller and shadier forests developed. An explosive, high-level radiation of plants happened in the Late Devonian, and the Mississippian is characterized by the continuing diversification of existing groups. All the major clades of plants were in existence by the mid-Mississippian: bryophytes (non-vascular plants), true ferns, lycopods (clubmosses), calamites (horsetails), and seed plants (represented by the pteridosperms or 'seed ferns'). Although today we primarily think of 'important' plants as being either gymnosperms (like conifers and cycads) or angiosperms (flowering plants), both of these groups are really just variations on the seed-plant theme that was pioneered by the pteridosperms.

These plants increasingly partitioned the Mississippian habitats, with forests of giant lycopods (e.g. *Lepidodendron*) in swamps, calamite trees along stream margins, and ferns and seed ferns in drier, upland habitats as a general rule. The vegetational structure of forests became modern during the Mississippian, with varying-sized trees in a stratified canopy, lianas (vines), epiphytes ('air plants' growing on tree trunks and branches), and ground cover. The complexity of animal-plant and animal trophic relationships lagged behind plant community structure; there were still no large herbivores. Plants supported the food chain by supplying abundant leaf litter that was fed upon by detritivorous insects and other arthropods, which in turn were food for each other and for the new kids on the block, the amphibians (some of which were also at least partly dependent on aquatic sources of food). As noted in the last chapter, winged insects had probably evolved by this time, but were not yet common elements of the fauna.

Chimaera shark *Echinochimaera meltoni*, Bear Gulch Limestone, Montana. Denver Museum of Nature & Science, Denver, Colorado.

As fossils of these more-complex plants appear, we are faced with an ongoing problem in paleobotanical taxonomy: the use of 'form genera' to designate plant parts, which may or may not reflect evolutionary relationships. Imagine a large tree (say, a pteridosperm) with leaves, trunk, roots, and seeds...unless you happen across an entire *in situ* fossil of the whole tree, how do you know which seed goes with which leaf? You can make an educated guess, but in fact, you don't really know for sure. Thus these separate plant parts are each given their own name. Different parts of one tree may be given several names, and many types of wood or foliage from different trees may be indistinguishable and lumped together in a single 'form genus.' It is almost never the case that the entire plant or tree is fossilized, partly because of dispersal (say, of seeds or spores) or abscission of leaves, and partly because parts tend to get broken up and jumbled in the fossilization process, especially if transport (like down a stream) is involved. This situation can only be resolved by finding linked parts as fossils, maybe a seed stuck to a leaf in one place and leaves stuck to the trunk in another. It's an ongoing problem, but the best that can be done at this point. Thus the plethora of plant names you run across in the literature can be misleading as to the true plant diversity in an ancient ecosystem. It also becomes more difficult to make reconstructions of these ancient plants. Nevertheless, compared with the Devonian, there was a clear increase in diversity of species in the Mississippian forests as well as in ecological complexity.

Blastoids *Pentremites godoni*, Ridenhower Formation, St. Clair, Illinois. Author's collection.

There was also an increase in diversity of freshwater animals, including the first freshwater clams, and an abundance of freshwater fish, including sharks. Predatory rhizodontiform fish (relatives of early amphibians that, like lungfish, could breathe air on occasion) grew to some six meters in length.

Paleozoic amphibians have traditionally been placed into two major groups: the labyrinthodonts and the lepospondyls. The labyrinthodonts are so named because of a labyrinthine infolding of the enamel which characterizes their teeth, and which the lepospondyls do not have. Labyrinthodonts share these characteristic teeth, fangs on the marginal bones of the palate, and a vertebral centrum composed of more than one element with the osteolepiform fish; they are clearly a monophyletic group. The lepospondyls, by contrast, consist of a heterogeneous assortment of groups which may have evolved separately from various labyrinthodont ancestors. Lepospondyls were characterized by simple teeth, a lack of palatal fangs, and a unitary vertebral centrum. We may cite a few characteristic members of these groups as representative of the diversity of Mississippian amphibians. The aïstopods were small (less than 50 cm/20 in long) lepospondyls, which were aquatic and possessed a streamlined body shape. The temnospondyls were labyrinthodonts that were about the same size (30-50 cm/12-20 in long) but were, in contrast, terrestrial insectivores. Finally, the anthracosaurs were a larger (up to 1.5 m/5 ft) group of labyrinthodonts which appear to have been semiaquatic carnivores, feeding on fish and maybe each other. Recently, an armored colosteid labyrinthodont was reported from western Kentucky that was covered from head to tail in half-inch-thick bony scales. Its eel-like shape and conical teeth indicate that it was an aquatic piscivore, whereas the armor attests to the hazards of living with the other, larger carnivorous amphibians and fish of the time. Amphibians clearly were at the top of the food chain in shallow-water and swamp habitats; however, with strong ties to aquatic environments and in the absence of herbivory, they had little impact on the plant communities of the time.

Crinoid *Taxocrinus colletti*, and coral *Cladochonus beecheri*, Edwardsville Formation, Indiana. St. Louis Science Center, St. Louis, Missouri.

Sometime during the Mississippian, the first reptile—a creature with an amniotic, or dry-land, egg—evolved. By the Early Pennsylvanian, a diversity of reptiles is represented, indicating an earlier origin. The erroneous notion that reptiles didn't appear until the Pennsylvanian has come into being because of a bias in the fossil record towards preserving animals (and plants) that lived in high-sedimentation (read: wet) environments. The fossil record of Mississippian tetrapods is not as complete as we would like, but they were clearly radiating into a variety of habitats. The details of this radiation, however, remain obscure. The first tetrapod which has been definitively identified as an amniote or reptile (based on the pattern of skull roof bones and tarsal configuration) comes from the mid-Mississippian of East Kirkton, Scotland, and looks fully terrestrial, having long, lightly-built limbs. This particular deposit is one of the few known from that time that represent drier, inland habitats. Mississippian tetrapod fossil localities are so few and far between that none are known from further than 5° north of, nor 20° south of, the paleoequator. Whereas it is possible that early tetrapods were restricted to tropical environments, it may be just as likely that other fossil localities at higher latitudes simply have yet to be found.

Crinoids, mixed species, Edwardsville Formation, Indiana. Private collection.

Where Mississippian Rocks Are Found

Limestones from the Mississippian Period are exposed widely in the Mississippi River valley and other areas primarily in the eastern half of the continent. They can be found underlying glacial deposits across most of Michigan's Lower Peninsula, save for the farthest north and southeast regions and a circular area in the center. There are also outcrops in central Indiana and Ohio; across much of central Kentucky and Tennessee on into northern Alabama; and in central Iowa. A quarter of the rocks exposed in Missouri are from the Mississippian, primarily in the northeast and southwest areas of the state, and extending into eastern Oklahoma and northwest Arkansas. A band of Mississippian limestone crops out along the Virginia-West Virginia border. Smaller outcrops can be found in the fault-block mountains of the basin and range province of Nevada, western Utah, and southeast Idaho. There's not much Mississippian rock in Canada, but there are a few outcrops in northern Nova Scotia and in the Canadian Rockies. The Brooks Range of Alaska also exposes some Mississippian sediments. South of the Rio Grande, Mississippian rocks are quite scarce.

Where Mississippian Fossils Are Found

Mississippian strata in Nova Scotia have yielded lungfish, amphibians, nautiloids, ostracods, arthropod trackways, gastropods, fish teeth, pelecypods, and *Lepidodendron* logs; however, following the distribution of Mississippian rocks noted above, most fossils from this time are found in the eastern and midwestern United States. Excellent Mississippian fossil collecting formations include the Edwardsville Formation of Indiana, which produces spectacular articulated crinoids, and the Newman Limestone, exposed in Tennessee and part of Kentucky, which yields a variety of marine invertebrate fossils. The world's longest cave system, Mammoth Cave in Kentucky, is formed in Mississippian limestones that produce bryozoans, brachiopods, crinoids, blastoids, and corals. North-central Arkansas' Imo Shale contains plentiful fossils of gastropods, cephalopods, bivalves, bryozoans, corals, brachiopods, crinoids, ophiuroids (brittlestars), and foraminifera; it hosts one of the richest gastropod/bivalve faunas known from this time period. This shale overlies the Pitkin Limestone, which is a crinozoan grainstone rich in bryozoans, crinoids, conodonts, foraminifera, bivalves, gastropods, and cephalopods. Alabama's Bangor Formation holds fossils of echinoderms, bryozoans, and corals. The Greenbriar Formation in Virginia and West Virginia is a series of oolitic and encrinoidal limestones that produces fossils of crinoids, corals, blastoids, and brachiopods.

Horn corals *Amplexizaphrentis pellaensis*, Pella Beds, Ste. Genevieve Formation, Oskaloosa, Iowa. Author's collection.

Missouri has several productive Mississippian formations: the Fern Glen Formation yields bryozoans, corals, brachiopods, pelecypods (clams), gastropods (snails), trilobites, blastoids, and shark's teeth; the Burlington Limestone produces horn corals, crinoids, brachiopods, and bryozoans (and extends into Arkansas and Iowa); the Warsaw Formation contains sponges, conularids, bryozoans, crinoids, and worm tubes; and the Salem Formation yields corals, brachiopods, and shark's teeth. Mississippian limestones and shales in eastern Oklahoma and the Arbuckle Mountains are good places to look for brachiopods, corals, crinoids, clams, and snails. Occasional plant fossils, including leaves, have been found in Oklahoma, but are rare from this time period. A fine assemblage of polyplacophorans (chitons, a type of mollusc) has been reported in the Humboldt Member of the Gilmore City Formation in Iowa. That state's McCraney Formation is rich in brachiopods, with lesser numbers of bryozoans, corals, crinoids, and clams; and the Wassonville Chert produces an abundant, similar fauna.

Probably the most amazing Mississippian lagerstätte in North America is the Bear Gulch Beds of Fergus County, Montana, which are in the Heath Formation of the Big Snowy Group. The fossil fish fauna from this locality, noted previously, is world-renowned. The Wood-hurst Member of the Lodgepole Limestone of eastern Idaho and western Montana yields a plethora of brachiopods, rugose and tabulate corals, crinoids, bryozoans, gastropods, bivalves, foraminifera, and conodonts. This fauna is very similar to that found in the Banff Formation of Alberta. Idaho's Arco Hills Formation is also known for its brachiopods, and shark's teeth can be found in the Scott Peak Formation. The Itkilyariak Formation of the Endicott Group, which outcrops extensively in the Brooks Range of Alaska, contains a marginal marine brachiopod-bryozoan fauna.

Several fossils including a trilobite glabella, a snail, & several brachiopods, all unidentified species. Sciotoville Bar Formation, Maysville, Kentucky. Author's collection.

Further south, brachiopods have been found in the Alamogordo Member of New Mexico's Lake Valley Formation. The Madison, Escabrosa, and Redwall limestones of northern Arizona produce some fine nautiloids, gastropods, crinoids, and occasional trilobites. In California's Inyo Mountains, the Chainman Shale has yielded many nice ammonoids.

Where Can Mississippian Fossils Be Seen in the Field?

Although not nearly as common as public preserves that showcase Devonian fossils, there are several places in North America where Mississippian fossils can be seen *in situ*. In the Midwest, our old friend Falls of the Ohio State Park at Clarksville, Indiana exhibits several Mississippian formations. The Edwardsville and

Ramp Creek formations show many examples of brachiopods, bryozoans, crinoids, and trace fossils. In the Beech Creek Limestone and Indian Springs Member of the Tar Springs Sandstone, you can see blastoids, brachiopods, bryozoans, horn corals, crinoids, clams, snails, and shark's teeth. This park is probably the best place on the continent to view Mississippian fossils in place.

Not too far away, if your dim-light eyesight is fairly good or you equip yourself with a strong flashlight, you may be able to spot some fossils embedded in the walls of Mammoth Cave. Located in south-central Kentucky near Cave City, the cave is carved out of the Mississippian limestones of the St. Louis, Ste. Genevieve, and Girkin formations. These marine beds contain fine examples of crinoids, blastoids, snails, brachiopods, horn corals, and even some shark's teeth. All tours of the cave are guided, however, so don't get so wrapped up in looking closely at the stone walls that you lose track of your group.

Unidentified nautiloid, Sciotoville Bar Formation, Maysville, Kentucky. Author's collection.

On the opposite side of the continent, the Tin Mountain Limestone exposed in California's Death Valley National Park contains brachiopods, corals, and crinoid stems. Remember, this is wilderness country, so make sure you are properly prepared. If you're an avid desert rat and in good physical shape, you may also want to hike into the Grand Canyon, where the Mississippian-age Redwall Limestone sports many examples of brachiopods, clams, snails, corals, and the rare remains of fish and trilobites. One advantage the Grand Canyon has over Death Valley is the abundance of tourists, so that you don't need to worry about hiking alone. Plus, even if you find the fossils to be scarce, you can't possibly complain about the scenery. And the Redwall Limestone is only about 1/3 of the way to the river, so you can easily hike in and out in a single day.

The end of the Mississippian is marked not by any mass extinctions or momentous events, but by the gradual emergence of much of the central continent above sea level and the transition to cooler and wetter climates of the Pennsylvanian—and the accompanying change in the rocks from primarily marine limestones to terrestrial coal swamps. It is to the flora and fauna of these coal swamps that we will next turn our attention.

Chapter Seven
The Pennsylvanian

At long last, most of our continent emerged from the waves. Swamps were everywhere — swampy forests of huge clubmoss trees, horsetails, and peculiar ferns with seeds. A large labyrinthodont amphibian basks on an Illinois shore, ignoring the gigantic millipede and dragonfly nearby; he's not hungry right now.

Pennsylvanian North America

When we began our look at the Paleozoic Era, our venue was solely the marine realm. Little by little, life gained a foothold on land, first colonizing permanently low, wet areas and only later venturing away from the waterfront. Plants pioneered; animals followed. By the time we reach the Pennsylvanian Period, or "Coal Age," complex forests were common in the fossil record, and those forests were teeming with animal life.

Definition and Nomenclature

Like the preceding Mississippian, the Pennsylvanian is considered a full-fledged geologic Period in North America, closely corresponding to what is known as the Upper Carboniferous in the rest of the world. This distinguishes the primarily marine limestones of the Mississippian from the predominantly terrestrial, frequently coal-bearing sediments of the Pennsylvanian, a difference which is far more apparent in North America than it is elsewhere. The name Pennsylvanian—after the state of Pennsylvania, where rocks of this age are abundantly exposed—was proposed by Henry Shaler Williams in 1891. Since 1906 the U.S. Geological Survey has considered both the Mississippian and Pennsylvanian to be Periods in their own right, whereas the International Commission on Stratigraphy only adopted the names in 2000, and then as Epochs (or subdivisions of the Carboniferous). Since our concern is the history of the North American continent, we shall herein consider the Pennsylvanian to be a *bona fide* Period.

The base of the Pennsylvanian, or Mississippian-Pennsylvanian boundary, has been defined since 1996 as the lowest occurrence of the conodont *Declinognathodus nodiliferus (sensu lato)* 82.9 meters above the top of the Battleship Wash Formation in Arrow Canyon, southern Nevada. The most recent dating places this boundary at 318 million years ago. The Pennsylvanian-Permian boundary, or end of the Pennsylvanian, was defined in 1996 at the lowest occurrence of the conodont *Streptognathodus isolatus* within the *S. "wabaunsensis"* conodont chronocline, 27 m above the base of 'bed 19,' Aidaralash Creek, Aktöbe, southern Ural Mountains, northern Kazakhstan. Currently this event is dated at 299 mya. Thus the Pennsylvanian may be said to span some 19 million years, making it the shortest time period in the Paleozoic Era.

Ferns, Pottsville Formation, St. Clair, Pennsylvania. Author's collection.

Paleogeography and Paleoclimate

You will remember that for much of the middle Paleozoic, North America had been tropically located and rotated somewhat clockwise from its present orientation. It continued its slow journey northward during the Mississippian, such that the Pennsylvanian equator can be located along a line running from present-day St. Louis, Missouri through New York City. The far northern reaches of Greenland, Alaska, and the Canadian Arctic isles were at long last beginning to emerge from the tropics into the temperate latitudes. Gondwanaland, too, rotated clockwise and was drifting northward—even faster than Laurasia (North America + Baltica + Siberia) was—and they were on a collision course, one that would eventually see the formation of the supercontinent of Pangea. As Africa nudged itself up against the southeast coast of North America, the present Appalachian Mountains were formed along the zone which had been previously rumpled by the Ordovician Taconic and Devonian Acadian orogenies. This—alternatively called the Appalachian or Alleghenian Orogeny—was the last phase of mountain-building to affect the eastern region of our continent before tectonic activity shifted westward. The nascent rumblings of this shift were seen in the formation of the Ancestral Rockies in Colorado, Utah, and New Mexico, although the seas continued to flood the continental shelf between these disconnected ranges. The Antler Mountains of Nevada became quiescent and were slowly wearing down. South America, still firmly attached to Africa's western edge, bumped into North America along what is now the Gulf of Mexico coast. Remnants of this collision can be seen in the Marathon Mountains of Texas, the Arbuckle and Ouachita Mountains of Oklahoma, and the Ozarks of Arkansas and Missouri. All of these disconnected hilly regions were once high mountains forming a chain continuous with the Appalachians.

Although it would not become apparent until Africa and South America separated from us (much later, during the Mesozoic), two very important, relatively small land areas were gained along North America's southern coast as part of the assembly of Pangea. The Florida peninsula is an African fragment that, during the Pennsylvanian, was wedged in at the triple-junction of the three modern continents. And you may recall that off the western coast of Peru were hung several small terranes that would form part of central Mexico as tectonic activity shifted to the western Cordillera.

Due to continuing glaciation on southern Gondwanaland and the reduction in atmospheric carbon dioxide effected by the spread of vegetation across the landscape, worldwide climates in the Pennsylvanian were cooler and wetter than Mississippian climates had been. In geological terms, the Pennsylvanian was an 'icehouse' world not all that different from today's—warm in the tropics but with a strong latitudinal gradient. North America, however, was warmer than it is now, by virtue of its being positioned entirely between latitudes 35°N (in Alaska) and 20°S (at the southern tip of Mexico). Most of the continent lay near sea level and played host to the vast 'coal swamps' from which the Carboniferous derives its name.

Lepidodendron leaves, St. Charles, Michigan. University of Michigan Exhibit Museum of Natural History, Ann Arbor.

A Coal Swamp Primer

In the modern world, the highest concentrations of vegetation are found in coastal wetlands such as the Dismal Swamp in North Carolina and Florida's Everglades. These are the nearest analogues we have for the Pennsylvanian coal swamps, although, as we shall see, the plant taxa inhabiting the swamplands some 300 million years ago were all very different from those we are familiar with today.

Swamps are officially called 'mires' and there are two main types: the 'planar mire,' typified by the Everglades, where the surface of the land is below the water table much of the year and standing water is abundant; and the 'domed mire,' like what is found in modern Sumatra, where the swampy environment and peat formation are maintained by abundant rainfall, even though the local water table is below the ground surface. Mires may have hummocky topography and hence wetter and drier areas within the swamp environment as a whole, or they may be featureless wetlands punctuated only by the relief provided by the standing vegetation. Plant matter accumulates faster than it can decay, and peat—perhaps eventually coal—forms. No doubt, both planar and domed mires could be found in the Pennsylvanian, depending on the topographic pavement and level of precipitation.

'Seed fern' seeds *Trigonocarpon*, Cedar Creek Co., Morrilton, Arkansas. University of Michigan Exhibit Museum of Natural History, Ann Arbor.

The presence of mires is a strong indicator of abundant rainfall in the Pennsylvanian tropics, possibly coupled with slightly cooler temperatures than equatorial latitudes had previously enjoyed.

Brachiopod *Spirifer rockymontanus*, Minturn Formation, McCoy, Colorado. Author's collection.

Let's take an imaginary trip into one of these Pennsylvanian coal swamps, but first, let's orient ourselves by starting in a place which will be familiar to many of you—Louisiana's Atchafalaya Swamp. Hopefully you didn't just whiz by on I-10 on your way from New Orleans to Houston, but got out and poked around a bit. Close your eyes. Try to ignore the gazillion mosquitoes that are trying to strip you to the bone like a school of voracious piranhas. Smell the air. It's so humid that it takes a bit of an effort to breathe. The atmosphere is dank, thick, with the earthy but not unpleasant smell of decaying vegetation. There's no breeze, just a close, sultry veil of warm mist that feels like it's on the verge of turning to steam. And the high-pitched buzzing—oh, the cacophony!—of myriads of insects everywhere. Open your eyes now. Before you stretches a bayou filled with water the color of black coffee. In the middle, it's open, but near the banks it's choked with water hyacinths and covered with the green scum of duckweed and algae. Bald cypress trees tower a hundred feet overhead, a closed-canopy forest in the water, their 'knees' rising two or three feet above the surface: that's how they breathe with their roots submerged. Some of them are lucky enough to have found some almost-dry ground on a low, emergent hummock. The water's edge is crowded with willow thickets—another tree tolerant of the mucky ground and abundant water. Out a ways, a white heron perches atop the snag left by a drowned cypress tree. In fact, there are dozens of such perches where the water is a little deeper—perfect for dragonflies. From the cypresses the Spanish moss hangs like curtains of flowstone, masking the mosses and lichens that grow like velvet on the damp, black bark.

What's that over by the rotting log in the shallows? At first it looked like another log, but now it slowly, sinuously slices a pathway through the duckweed... it's an alligator! He's not humongous, maybe only four or five feet long, but still, you don't want to mess with him. Better to stay on the shore and leave him be. The place where he disappeared into the water hyacinths holds your gaze for a long time, but he doesn't reappear. Now there's a strange, scratching noise behind you, and a masked face pokes 'round a large tree. Just a curious raccoon. Charming, really. But something's bothering your toes. You wiggle them, but it doesn't help, so you glance down at your sandal. Ugh!—and a reflexive kick. A cockroach! No, thanks!

Walchia, an early conifer, Minturn Formation, McCoy, Colorado. Author's collection.

Ok, now close your eyes again.

A Pennsylvanian 'Flashback'

The air is sultry, thick, and close, still so humid that your breathing is a little hindered. It's so warm that it reminds you of a steam bath. The insect chorus is unchanged, but thank heavens, the mosquitoes are gone. And the tickling at your toes refuses to cease. That dratted cockroach, you thought you had gotten rid of him, so you open your eyes to see what the problem is. Horrors! It's a cockroach, all right, but this one is four inches long! You kick again, and he goes careening off into the distance, past a fallen log. Good riddance! My, that's a strange fallen log. Its bark is all in a criss-cross, diamond pattern, and it's a beautiful shade of emerald green. Parts of it are cloaked in velvety green moss and are sprouting delicate shelf fungi. You've never seen a log like that before. Slowly you look around you. You're obviously still in the swamp—over there is the bayou with the black-coffee water—but instead of standing in a shady, closed-canopy forest, here the woods are more open, and sunlight reaches all the way to the ground. That accounts for the green trunks (and even green roots!) of the trees: they're photosynthetic too! You look up. Most of these trees don't even have crowns; they look like verdant, diamondback telephone poles. A few, though, sport umbrella-like crowns at their apices some hundred feet or so above you. The foliage is lacelike and repeatingly forking...it reminds you of the 'ground pine' that you've seen growing inconspicuously on the floor of the oak-hickory forest back home. You try to remember...ground pine is some sort of primitive plant... oh, yes, it's a clubmoss...and these trees are clubmosses too, only gigantic ones! Obviously, they only sprout those crowns when they're ready to reproduce, spending most of their lives as unbranched poles. *Lepidodendron*, they're called, and in this Pennsylvanian coal swamp they are the dominant forest trees on the emergent ground close to the water level.

Calamite leaves *Annularia radiata*, Mazon Creek, Kankakee Co., Illinois. Private collection.

They aren't the only trees, though. Out in the shallows, where the cypresses encroached in the Atchafalaya, are gigantic horsetail-trees called *Calamites*. You're sure they're horsetails because of the jointed trunks, which look exactly like the knee-high scouring rushes you're familiar with. You wonder if they'd separate easily at those nodes like the ones you played with as a kid, but maybe it's better not to try, and risk a fifty-foot tree coming crashing down on you! There seem to be a couple of kinds, those which have circumferential rings of fringelike leaves emerging from

Lepidodendron bark, St. Charles, Michigan. University of Michigan Exhibit Museum of Natural History, Ann Arbor.

the joints, and those which don't. They seem to like growing right in the shallow water, where it's covered with algae and small floating plants that you can't identify. Their stems are green too, like the lycopods'. In fact, everything is green here, every plant surface dedicated to catching sunlight and making food. Out farther in the water are some drowned lycopod stumps, identifiable by that pineapple-like bark. No herons perching on them here, but what's that thing with the two-foot, lace-like wings? Why, it's a gigantic dragonfly, and except for its size, you wouldn't know it from one back home. Its body shimmers with a blue-and-gold iridescence; you gaze at its beauty for several minutes, until a movement in the water-scum catches your eye. That's no alligator...its skin is moist and bumpy, and its head is flat and broad. It looks like something you might get if you crossed a toad with an alligator. It seems lethargic enough until it bolts halfway out of the water to catch one of those huge dragonflies in its sieve of tiny, sharp teeth. An anthracosaur, one of those labyrinthodont amphibians that gave rise to the reptiles, and ultimately, to us. You decide that he's one who's better left alone.

Your gaze shifts back into the forest. With all that sunlight reaching the ground, it sports a tangled mantle of what are obviously ferns. None of them are familiar species, but you're sure they're ferns nonetheless. There are no curtains of Spanish moss here—it's a bromeliad, a flowering plant, one of a group that won't make its debut for another hundred and fifty million years. But the forest isn't merely made up of a carpet of ferns punctuated by stands of scaly telephone poles. Vines with fernlike foliage wrap themselves around the lycopod trunks and loop in scallops between them. Now, those are some really strange ferns: every so often, along the frond rachis, one of the pinnae is replaced by a nutlike seed the size of a pecan. These are pteridosperms, commonly called 'seed ferns,' and they are the forerunners of all the seed plants we know today. Not all seed ferns are herbaceous, though—over there, atop that hummock, where the soil isn't water-logged, are some of their cousins with a tree-like habit, called *Medullosa*. If it weren't for those nutlike seeds, you might confuse them with the tree ferns that grow in the moist forests of New Zealand today.

Cordaites principalis leaf, Topeka Limestone, Kansas. Denver Museum of Nature & Science, Denver, Colorado.

But there's a scrambling noise in the ferns off to your right. Obviously, you're not alone here. Suddenly, a grey, scaly head pops up above the verdant fronds. It's a captorhinid reptile, vaguely reminiscent of a modern monitor lizard, and in his jaws he holds a struggling, fluorescent orange, two-foot millipede. Well, you think, if I had to eat something around here, better that than those cockroaches! A second later, the head is gone, and all that reveals his presence is a rustling noise and

Amphibian *Saurerpeton obtusum*, Linton Formation, Ohio. Denver Museum of Nature & Science, Denver, Colorado.

motion among the ferns that indicates he's off after his next snack. You close your eyes again, feeling the gentle caress of a wisp of breeze and listening to the symphony of insect wings. No airplanes roar overhead, no birds squawk, and best of all, no mosquitos interrupt your reverie. You think you'll stay here for awhile, because when you open your eyes once more, all that will be left of this place is seams of coal and misty memories.

Life in the Pennsylvanian

This scenario could have taken place in innumerable locations across North America during the Early or Middle Pennsylvanian. During this time, the giant lycopod trees comprised about 70% of the arborescent vegetation in the coal-swamp forests. We know this by how they dominate the fossil remains found in coal seams. Unimaginable numbers of these trees went into the creation of the peat and, later, coal deposits which underlie the eastern regions of North America. It has been estimated that some tens of meters of vegetation went into the making of a single meter of coal. The vast majority of coals worldwide were formed during the Pennsylvanian; on our continent, these coals are found primarily in the east, in four main areas: the central Michigan Basin, the Illinois Basin, the western flank of the Appalachians from Pennsylvania to Alabama, and in the Midcontinent Basin which stretches from Iowa to Oklahoma.

As we have seen, a typical early-to-mid-Pennsylvanian coal swamp was dominated by lycopod 'pole trees,' which sprouted parasol-like crowns of leaves towards the end of their lives, when they were ready to reproduce, and then died. They formed relatively open forests with an understory of ferns and seed ferns, some of which had a vinelike or epiphytic habit. Giant horsetails were abundant fringing the shallow water and in disturbed areas, where they could reproduce clonally via rhizomes. Pteridosperm trees and tree ferns preferred moist but better-drained soils either at elevation within the bogs or near their edges. Drier, upland areas were dominated by cordaite trees, which were gymnosperms (non-flowering seed plants) with long, straplike leaves, and the earliest confers, typified by the form genus *Walchia*, which is reconstructed as somewhat resembling a modern Norfolk Island Pine. Although conifers were restricted to the upland communities, cordaites had a wider environmental tolerance and may have been found in intermittent stands in the coal swamps as well, where a few species are even thought to have had a mangrove-like habit. As during the Mississippian, an absence of growth rings in these Pennsylvanian trees attests to a uniformly warm, tropical climate for our continent at the time.

Fern *Alethopteris serlii*, Francis Creek Shale, Illinois. Burpee Museum of Natural History, Rockford, Illinois.

Pennsylvanian sediments throughout the world are cyclic in nature, probably because of the sea level changes brought about by the waxing and waning of the glaciers on Gondwanaland. A typical sequence, or cyclothem, consists of a basal sandstone recording an ancient beach; siltstone representing a tidal flat; a fresh-water limestone formed in a shallow lagoon; a terrestrially-derived clay layer, which was the soil on which the coal forest became established and which commonly preserves root traces; coal that was formed from the compacted peat laid down while the coal swamp flourished; shale representing nearshore tidal deposits; a shallow marine limestone; and finally a deep marine black shale. This sequence of layers may be repeated as many as 50 times within a given formation. Such cyclicity occurs in coastal sediments when the seashore transgresses and regresses (moves in and out) fairly quickly and in a repetitive fashion.

The boundary between the Middle and Late Pennsylvanian was marked by a drying trend, although this just meant going from very rainy to moderately so. Coal swamps persisted but were not as widespread as they had been. There was, however, a profound change in their floral makeup. All of the lycopod trees except the single genus *Sigillaria* became extinct, and they were largely replaced by forests of tree ferns, cordaites, and seed ferns. One of the most famous fossil assemblages known from the Late Pennsylvanian is the Mazon Creek biota of Illinois. At the time, the Illinois Basin was a shallow gulf into which spilled a large river and delta complex from the northeast. You can imagine something similar to the Mississippi River Delta with a tangle of freshwater and brackish distributaries separated by tidal flats and mud bars on which grew the coal swamp vegetation. Mazon Creek fossils are found in concretions ('coal balls') of siderite (iron carbonate), which formed when plants and animals fell into the shallow water and were fairly rapidly covered with sediment. Initial decomposition of the organism produced carbon dioxide which combined with iron in the sediments to form iron carbonate. The formation of ironstone entombed the organism and arrested further decay, leading to the amazingly detailed preservation for which the Mazon Creek nodules are famous. Even soft-bodied animals, like the "Tully monster"—a carnivorous wormlike organism of uncertain taxonomic affinity—were preserved. The fact that siderite is a harder mineral than the surrounding sediments of the Francis Creek Shale makes collecting of these concretions easy. They weather out of the shale in mining spoil piles and can be cracked with a hammer along the plane of weakness containing the fossil. About one in ten concretions harbors a fossil.

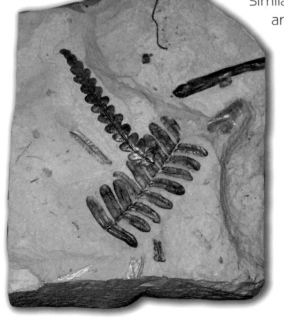

Fern *Alethopteris grandini*, Tonganoxie Sandstone Member, Stranger Formation, Franklin Co, Kansas. Private collection.

The Mazon Creek is considered a lagerstätte and is probably the best-known Pennsylvanian biota in North America, both because of the number and exquisite preservation of the fossils, and because many of them represent species found nowhere else in the world. The Mazon Creek assemblage can be divided into two distinct ecosystems: the Braidwood biota, which represents a typical Late Pennsylvanian coal swamp, and the Essex biota, which reveals a shallow marginal-marine environment. The Braidwood biota preserves mostly seed ferns, calamites, and fern foliage, with some insects, myriapods, freshwater fish (including the only known fossil hagfish), clams, and ostracods, and with plants outnumbering animals about 100:1. The Essex biota is divided about fifty-fifty into plant and animal fossils and preserves the remains of jellyfish, worms, snails, clams, shrimp, fish, and the Tully monster, among others. Together with the slightly younger Calhoun Coal of the Mattoon Formation, these Mazon Creek fossils yield the best-studied Pennsylvanian insect fauna in the world and a unique glimpse into the coal swamps of the time.

Tully Monster *Tullimonstrum gregarium*, Francis Creek Shale, Illinois. Burpee Museum of Natural History, Rockford, Illinois.

Most groups of amphibians which had appeared during the preceding Mississippian persisted into the Pennsylvanian, including the very successful labyrinthodonts and their newly-evolved branch, the eryopoids. The earliest-known fossil eggs are from the lower Permian, although as we have seen, osteological evidence tells us that reptiles with an amniotic (dry-land) egg had also gotten their start in drier inland areas during the Mississippian. An abundance of wetter environments allowed for the better preservation of the tetrapod fossil record during the Pennsylvanian. The oldest reptile fossil in North America is an Early Pennsylvanian trackway from eastern Kentucky, but one of the earliest well-preserved terrestrial tetrapod faunas comes from near Joggins, Nova Scotia and is only slightly later in time. There, good skeletons of reptiles and amphibians have been found in the infillings of lycopod tree trunks. Lycopod trees had only a sparse layer of woody tissue and gained most of their support from their thick bark. When the coal swamps were occasionally flooded, these pole trees drowned and their roots were buried several feet deep in sediment. The interiors rotted away, leaving hollow, buried stumps ringed by the thick bark. During drier intervals, these were left as pothole-traps, and unwary tetrapods fell in and died (they may have drowned in standing water inside the stumps). Sediments later infilled the trees, preserving the bones.

Seed fern *Neuropteris* sp., Monroe Co., Ohio. Private collection.

During the Pennsylvanian, the reptile family tree split into three branches which were to be of great evolutionary import during the Mesozoic: the anapsids, the diapsids, and the synapsids. These three groups are distinguished based on the pattern of postorbital fenestrae (holes in the skull behind the eye socket). The anapsids lacked any fenestrae and are represented today solely by the living turtles. The synapsids possessed a single fenestra low on the skull, bordered by the squamosal, jugal, and postorbital bones. This group includes the Permian mammal-like reptiles (or protomammals), as well as the mammals to which they would subsequently give rise. The diapsids possessed a fenestra identical to that found in synapsids, but also a second one above the squamosal, lateral to the parietal bone. One lineage of diapsids led to modern lizards and snakes as well as to the marine plesiosaurs, ichthyosaurs, and mosasaurs. The other diapsid branch spawned dinosaurs, crocodylians, and eventually birds. Although the traditional Class Reptilia cannot be rigorously defined in modern cladistic terms—except to say that a 'reptile' is an amniote that isn't a mammal or a bird—in common parlance the concept of a reptile is widely understood and a useful one, so we will retain it here.

Three skull types (see text): Anapsid (top) - sea turtle *Protostega*, Cretaceous; synapsid (middle) - pelycosaur *Dimetrodon*, Permian; diapsid (bottom) - dinosaur *Allosaurus*, Jurassic. From displays at Denver Museum of Nature & Science, Denver, Colorado.

While both plants and animals were diversifying at a dizzying pace, the interactions between them took some time to develop. During the Early Pennsylvanian, the main route for plant productivity into the food chain was still via detritivorous arthropods. Insect diversity was increasing, though, and at least 10 orders were present by the Early Pennsylvanian. There is tantalizing evidence for flight in insects as far back as the Late Devonian, but it was surely well-developed by the Pennsylvanian with the whole-body preservation of insects with wings. Most of these, like the giant dragonflies, however, were primitive types that could not fold their wings back over their bodies. Flight may have been an impetus towards the development of herbivory in insects, as it made high-energy food resources (like seeds and pollen) found high up in the forest canopy more accessible than they had been heretofore. By this time, all methods of insect herbivory except leaf mining (folivory, root-feeding, sapsucking, consumption of pollen grains, and boreholes in seeds) may be observed in the fossil record. The first suggestion of tetrapod herbivory is seen in the dentition of the Late Pennsylvanian diadectid cotylosaurs, whose teeth are differentiated into sharp, chisel-shaped anterior 'incisors' and blunter cheek teeth showing wear facets and suitable for chewing plant matter. These amphibians may have been omnivorous, supplementing a diet of insects and millipedes with seeds or a salad now and then. The first freshwater clams and land snails appeared at this time and may also have made tasty morsels. There weren't any totally herbivorous tetrapods yet, however, and therefore overall ecosystem complexity did not yet rival that of subsequent times.

'Modern' complexity of plant ecosystems preceded that of animal interactions by some tens of millions of years. Terrestrial animals during the Pennsylvanian can be divided into four trophic (food) guilds: detritivores, which were mostly arthropods; herbivore-omnivores, many arthropods plus the aforementioned diadectids; insectivores, the larger arthropods and the smaller tetrapods; and carnivores, which were primarily large tetrapods. Thus, the amount of energy transmitted into the food web via herbivory would remain small until the widespread appearance of tetrapod herbivory in the Permian. In other ways, though, the Pennsylvanian coal swamps were even more diverse than modern ones—at no other time in history were the forests home to more separate orders of trees. In fact, Pennsylvanian plants are the best-known of the entire Paleozoic because of their preservation in coal swamp sediments, particularly in 'coal balls.'

If up until now we have neglected to mention life in the epicontinental seas, it is because this differed little from that of the preceding Mississippian. Reef-building organisms still had not recovered from the Late Devonian extinction, although rugose and tabulate corals were still around in lesser numbers. Forests of crinoids still decorated the topographic highs on the seafloor, and brachiopods, nautiloids, ammonoids, and conodonts were common. Trilobites were in serious trouble with only a handful of species still surviving. Fish continued to diversify and reign as top predators, especially the sharks. Fusulinid forams became so common that in some formations their tests form a major proportion of the rock volume.

Where Pennsylvanian Rocks Are Found

As we have seen, the largest exposures of Pennsylvanian rocks in North America are in the eastern United States, in areas that were shallow basins during the late Paleozoic. Probably the most extensive of these is the Midcontinent Basin, where Pennsylvanian rocks outcrop in south-central Iowa and the far southeastern corner of Nebraska, across most of the northwestern third of Missouri, in eastern Kansas and Oklahoma, and into west-central Arkansas. Smaller, disconnected outcrops extend south into central Texas. Pennsylvanian sediments deposited in the Illinois Basin underlie all but the furthest northern regions of that state and continue into southwestern Indiana and western Kentucky. The Cincinnati Arch separates the Pennsylvanian rocks of the Illinois Basin from those in the Appalachian Basin to the east. These stretch from western Pennsylvania, eastern Ohio, and most of West Virginia southward through eastern Kentucky and Tennessee into north-central Alabama. The Michigan Basin was filling in and hosts Pennsylvanian bedrock only in its central region, which encompasses the interior regions of the Lower Peninsula. Many of these rocks are inaccessible due to a mantle of glacial till. Isolated Pennsylvanian outcrops may be found in the mountainous

Tree trunk Sigillaria, Palo Alto, Pennsylvania. University of Michigan Exhibit Museum of Natural History, Ann Arbor.

regions of British Columbia, the Yukon, and Alaska, as well as on the Canadian Arctic islands of Cameron and Cornwalis, and the western shore of Greenland. Small exposures are found in the eastern Canadian provinces of New Brunswick and Nova Scotia, associated with coal fields that extend eastward under the Atlantic seafloor. Pennsylvanian outcrops are scarce in the western states, but can be found in some of the mountainous regions of Colorado, Utah, and Arizona, especially at the bottoms of deep canyons. Alas, south of the border, there is a dearth of rocks from all of the Paleozoic.

Where Pennsylvanian Fossils Are Found

We have already mentioned Illinois' Francis Creek Shale of the Carbondale Formation and the Calhoun Coal of the Mattoon Formation as hosts to the siderite concretions containing a plethora of well-preserved Pennsylvanian fossils. The Carbondale Formation extends from Indiana and western Kentucky into Oklahoma, where its Colchester Coal and Mecca Quarry Shale members also are highly fossiliferous, the former yielding a coal-swamp biota and the latter producing nice brachiopods and fish. A Mazon Creek-like (Braidwood) flora and fauna is also found in the Carbondale's Energy Shale Member in southern Illinois. The tetrapod and lycopod fossils of Nova Scotia's Joggins Formation have also been touched upon, whereas calamite and seed fern foliage may also be collected there; nearby, New Brunswick's Pictou Formation has been productive of many plant leaf fossils. Pennsylvania's Llewellyn Formation preserves many nice *Lepidodendron* trunks as well as the foliage of seed ferns, and is in fact a classic seed-fern unit. Eastern Kentucky's Breathitt Group is very rich in fossils. This is a typical sequence of cyclic sediments wherein the Magoffin and Kendrick Shale members of the Four Corners Formation yield a brachiopod-dominated assemblage which also contains nautiloids, ammonoids, crinoids, snails, and clams; elsewhere in the same formation, in terrestrial sediments associated with coal seams, beautiful calamite trunks and foliage, *Lepidodendron* trunks, stigmarian root casts, and seed fern and true fern leaves can be found. The Conemaugh Formation outcrops in Ohio, West Virginia, and northeastern Kentucky, where it contains brachiopods, crinoids, gastropods, and fusulinids. Western Kentucky's Tradewater Formation produces abundant crinoids, bryozoans, corals, and fusulinids, particularly in the Curlew Limestone Member. West Virginia is almost entirely underlain by Pennsylvanian rocks. The Kanawha Formation, in particular, yields a prolific variety of plant fossils including *Lepidodendron, Sigillaria, Cordaites, Calamites,* and various seed ferns and true ferns. Missouri and Iowa's Altamont Formation, most

Ferns, Pottsville Formation, St. Clair, Pennsylvania. Private collection.

notably the Lake Neosho Shale, produces an abundance of fossils including sponges, bryozoans, tabulate corals, brachiopods, various molluscs, trilobites, crinoids,
and shark's teeth. Crinoids are also abundant in limestones of Missouri's
Cherokee Group, as they are in Illinois' LaSalle Formation. In the farthest
southeast corner of Nebraska, marginal marine crinoid-brachiopod-coral
faunas are found in the Dennis, Oread, Lecompton, and Stanton formations. The Topeka Limestone of eastern Kansas is known for a diverse ecosystem made up of lycopods, cordaites, seed ferns, walchian
conifers, calamites, fish, eurypterids, amphibians, reptiles, scorpions,
millipedes, dragonflies, and crickets. Kansas' Stanton Limestone also produces a spectacular insect fauna, and other formations in the eastern part
of the state can be collected for lycopod, calamite, and seed fern fossils
(both trunks and foliage). Tetrapod fossils have been reported from the
Ada Formation of Oklahoma. Pennsylvanian outcrops in northern Texas
represent some of the most fossiliferous marine shelf sediments known
from this time, including the Lake Bridgeport Shale member of the Wolf
Mountain Formation's molluscan and sponge-crinoid faunas (with rare trilobites) and the sharks, corals, and cephalopods of the Finis Shale. The nearby
Saddle Creek Limestone also produces superb shark remains. The Minturn Formation of central Colorado is a great place to collect crinoids, brachiopods, shark's
teeth and spines, molluscs, bryozoans, rugose corals, and plant fossils including
cordaites and walchian conifers. This fauna is similar to what is found in the Honaker Trail Formation of eastern Utah. That same state's Manning Canyon Formation
yields a beautiful array of plant leaf fossils. The Supai Formation of Arizona also
produces many fine ferns, conifers, and horsetails, and the Naco Formation yields
brachiopods, clams, and horn corals.

Trilobite *Ditomopyge
scitula*, Allegheny
Formation, Lower
Kittanning Member,
Baltic, Ohio. Private
collection.

Where Can Pennsylvanian Fossils Be Seen in the Field?

It is perhaps surprising, with so many spectacular Pennsylvanian sediments
exposed in the eastern U.S., that so few preserves have been erected to showcase
fossils from this timeframe. Fortunately, easterners have only to travel a short way
north of the line to visit Cape Chignecto Provincial Park near Joggins, Nova Scotia,
which protects a world-famous fossil site for Pennsylvanian vertebrates and plants.
Here, exposures of the Joggins Formation, especially those in cliffs by Chignecto
Bay, preserve *in situ* (in place) standing tree trunks, mostly those of the lycopods
Lepidodendron and *Sigillaria*. Inside some of these hollow stumps have been found
complete skeletons of several varieties of terrestrial tetrapods, leading scientists
and collectors to wonder whether hollow stumps were a preferred shelter for these
animals or held standing water into which they fell to their doom.

Ironically, the most famous Pennsylvanian fossil-bearing unit in all of North
America—the Francis Creek Shale with the Mazon Creek nodules—can claim no

special protected locality as its own. Visitors to northern Illinois wishing to see Mazon Creek fossils outside of a museum should ask around in the Chicago area for access to strip mining dumps located on private land. The Earth Science Club of Northern Illinois (ESCONI) may be your best resource for information on this score.

Fern *Alethopteris* sp., Henry Co., Missouri. University of Michigan Exhibit Museum of Natural History, Ann Arbor.

The western states harbor more opportunities to see Pennsylvanian fossils in the field, although these are mainly shallow marine sediments and not the remains of coal swamps. In the Grand Canyon of northern Arizona, the Supai Group grades from muddy shales in the eastern region of the canyon (which contain numerous fossils of plants, amphibians, and reptiles from an ancient river delta) into marine limestones and dolomites in the western canyon. The limestones contain many fine examples of marine invertebrate fossils.

Probably the most prolific, and widely-exposed, unit for Pennsylvanian fossils in the Western Interior is the Honaker Trail Formation. Best-represented along (where else?) Honaker Trail at Goosenecks of the San Juan State Park in southeastern Utah, this grey marine unit contains an abundant and diverse marine invertebrate fauna including many corals, bryozoans, brachiopods, and crinoid stems. This same formation also crops out in Canyonlands National Park, mostly in deep canyons along the Colorado River, and near the visitor's center for Arches National Park near Moab, Utah.

The assembly of Pangea, which continued into the Permian, brought changes in atmospheric circulation and precipitation that spelled doom for the coal swamps of North America. Part of the reason for this was the slow, inexorable march of Laurasia toward the subtropical dry belt around 30°N, which drastically attenuated the rainfall our continent received. Additionally, strong monsoons were generated by the enormous landmass of the supercontinent, but the heavy seasonal rains were insufficient to maintain the moisture level needed by the mires year-round. Spore-bearing plants like the calamites and lycopods lost ground to the seed plants; and even the pteridosperms were largely replaced by the more dry-adapted conifers and cycads.

The Pennsylvanian had seen the development of the first truly complex forests. Although they are gone forever, the legacy they left us—a rich source of energy and an even richer fossil record—still touches us some 300 million years later.

Chapter Eight
The Permian

The Permian saw North America joined with all of the other continents, forming one gigantic landmass, Pangea. The climate was hot and dry, with strong monsoons. This New Mexico sunset shows a fin-backed pelycosaur, *Dimetrodon*, hungrily eyeing the boomerang-headed amphibian *Diplocaulus*. But their days were numbered — the end of the period was marked by the greatest mass extinction of all time.

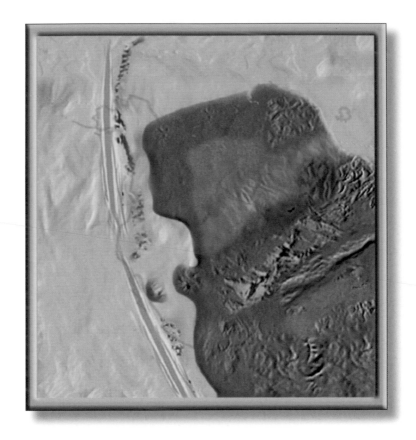

Permian North America

Now we come to the Permian, the final period in the Paleozoic Era. Early geologists had recognized that there were major changes in the types of fossils represented during three distinct time-spans of the Phanerozoic Eon ("time of revealed life"), and, on this basis, defined the Paleozoic ("ancient life"), Mesozoic ("middle life"), and Cenozoic ("recent life") Eras. As geological and paleontological knowledge grew, each of these was subdivided into several periods, as we have seen. It will become evident with our discussion of the Permian why the end of this period also marks the end of an era.

Definition and Nomenclature

The name Permian was given to the last period of the Paleozoic by English geologist Sir Roderick Murchison in 1841, after he had travelled to Russia to study the rocks there. He chose the name in honor of the ancient kingdom of Permia and its namesake, the Russian city of Perm, situated on the west flank of the Ural Mountains, where rocks of this age are extensively exposed.

The Permian follows the Pennsylvanian, and this boundary is defined as the lowest occurrence of the conodont *Streptognathodus isolatus* within the *S. "waubaunsensis"* conodont chronocline, 27 m above the base of 'bed 19,' Aidaralash Creek, Aktöbe, southern Ural Mountains, northern Kazakhstan. Currently this boundary is dated at 299 mya. The Permo-Triassic boundary, or end of the Permian, was defined in 2001 at the lowest occurrence of the conodont *Hindeodus parvus* and the termination of a major negative carbon isotope excursion at the base of 'bed 27c,' Meishan, Zhejiang, China. The most recent dating of this event is 251 mya, such that the Permian encompasses some 48 million years.

Paleogeography and Paleoclimate

The continents at this time were assembling themselves into one giant supercontinent, Pangea, which stretched almost from pole to pole. Antarctica still straddled the South Pole and Siberia reached to above the Arctic Circle. You will recall that during the past billion years, two other supercontinents, Rodinia and later Pannotia, had previously encompassed essentially all the world's lands. At other times, what are now the northern continents danced about the globe, while the southern landmass of Gondwana remained largely intact, embracing South America, Africa, Australia, India, and Antarctica. During the Pennsylvanian the northern and southern continents became reunited. The crustal movements that created Pangea were largely completed by the Early Permian, so that most of the period was tectonically quiescent. A vast ocean, the Panthalassic (forerunner of the modern Pacific), occupied more than half the globe. Pangea itself was shaped much like a pie with a piece missing—the triangular gap facing east and comprising a smaller seaway, the Tethys, which separated east Gondwana from Eurasia. North America was largely in the northern hemisphere by this time, with the equator running from central Texas through northeastern North Carolina. With the suturing of Africa to our east coast and of South America to our southern, North America gained the last two significant continental pieces to be added to our land area—much of Mexico which, although somewhat uncertain in its origin, is thought to have previously been a South American fragment hanging off the Peruvian coast; and Florida, which had been a small piece of Gondwana lying between West Africa and northern Brazil. Since that time, the only lands to be added to the North American continent have been various small terranes accreted along the western (Cordilleran) margin.

Monumental changes in global climates were effected by the assembly of Pangea. We have previously noted the beginning of a drying trend that commenced in the Late Pennsylvanian, causing a reduction in the area of the 'coal swamps.' With the onset of the Permian, large areas of the continents became

Permian reef fragment with brachiopods *Collemataria elongata*, *Meekella* sp., *Grandaurispina* sp., & *Paucispirifera* sp., Gaptank Formation, Texas. Denver Museum of Nature & Science, Denver, Colorado.

truly arid at the expense of tropical forests. Most of what is now the United States became a desert. Coastal area and maritime influence were substantially reduced when so much land was agglomerated together, a likely factor in this extensive continental dry-out. The large landmass also generated strong monsoons which brought extremes of rain and drought to many areas. Wet-adapted floras, like those of the Pennsylvanian, could not survive such seasonality of rainfall; they largely vanished from North America (and indeed, most of the world), although some coal swamps hung on in China into the Permian, as this was an equatorial peninsula in eastern Pangea with more reliable access to maritime moisture.

Along with the drying trend was an overall warming as the glaciers on Gondwanaland waned and finally disappeared altogether. This was probably due to a combination of factors. For one, by this time only part of Antarctica still covered the South Pole; much of the rest of Gondwanaland had moved northward into more equable climes. Another factor may have been the pole-to-pole reach of the Panthalassic Ocean, in which large circulatory gyres could bring warm equatorial waters to polar latitudes uninterrupted by deflecting landmasses. Although cool temperate, and even cold, climates prevailed at high latitudes in both hemispheres, by the mid-Permian the icecaps were gone. In the interior regions of tropical Pangea, the lack of maritime influence allowed baking temperatures to be reached in the deserts, just as we have in the Sahara and Australian Outback today.

Life in the Permian

The changes that were so detrimental to spore-bearing plants like the lycopod (clubmoss) and sphenopsid (horsetail) trees gave a distinct advantage to the seed plants. They could reproduce without a source of standing water, and rapidly took over the drier habitats. Cordaite trees, more tolerant of water stress than other plants which had been common during the Pennsylvanian, hung on throughout the Permian but became extinct at the end of the period. The primitive walchian conifers gave rise to the first modern conifers during the Early Permian, and these quickly came to dominate the upland forests. The Gnetales, an enigmatic group of gymnosperms that contains the modern *Ephedra* (Mormon tea), first appeared during the middle Permian of North America. From pteridosperm stock arose a new gymnosperm group, the cycads, which was to become increasingly successful during the Mesozoic. They are thought to have evolved from the medullosan group of seed ferns because these (unlike other pteridosperms) bore separate sterile and fertile fronds. In cycads the fertile leaves are grouped in cones borne at the apex of a stout trunk and encircled by a crown of fernlike or palmlike fronds. Both the cycads and the conifers were far better adapted to arid habitats than typical Pennsylvanian plants had been, although a few seed ferns hung on in moister areas even into the mid-Mesozoic. Permian forests therefore took on a far more modern aspect than we have thus far seen.

With the melting of the glacial ice came a global rise in sea level that flooded the low-lying areas of the continents. Most of North America remained above water, but areas of Mexico, Arizona, New Mexico, Texas, and the Great Basin states played host to shallow seas. Extensive sponge-bryozoan-algal reefs developed in west Texas and southeast New Mexico; the Capitan Reef there was a vast complex perhaps comparable to modern Australia's Great Barrier Reef. In the shallow back-reef evaporating ponds, thick sequences of salt and gypsum were laid down. More salt deposits are known worldwide from the Permian than from any other time in Earth's history. The Permian limestones of the Capitan Reef are extremely fossiliferous and include the Carlsbad Limestone in which the Caverns of the same name are formed.

Life in the seas differed little from that of the preceding periods of the Late Paleozoic. Fusulinid forams were common and in some Texas formations make up 20% of the rock. Brachiopods, bryozoans, and ammonoids thrived; trilobites were rare. Clams and snails were doing fine, thank you. Crinoids remained common, and bony fish flourished. Tabulate and rugose corals contributed to the reef-building fauna.

But the real faunal changes were to be seen on the land. With the reduction in swamp habitats, amphibians lost major ground to the reptiles. Although it is thought that the amniote egg probably evolved during the Late Mississippian or Early Pennsylvanian, the first known hard-shelled fossil egg is from the Lower Permian of Texas. The lag may be due to a dearth of preservation in earlier upland habitats; alternatively, it may only have been in Permian times that the amniote egg evolved a fossilizable hard shell. Some modern salamanders lay their eggs in moist terrestrial areas such as rotten logs, and the young hatch out as miniature adults without going through a larval stage. It is reasonable to assume that this is how the amniote egg got its start, and it may have taken some time before it developed a rigid shell and was completely ready for a truly dry-land environment.

Who were these reptiles that took over the face of the continents? In the last chapter, we met the three lineages designated the Anapsida, Synapsida, and Diapsida. Each of these spawned a variety of groups during the Permian. Among the anapsids—those reptiles lacking any postorbital fenestrae (or holes in the skull behind the eye socket)—an assortment

Ferns & conifer leaves, Abo Formation, Las Cruces, New Mexico. New Mexico Museum of Natural History, Albuquerque.

of primitive groups is seen, most of which failed to survive beyond the end of the Permian. The turtles, the only anapsids still living, do not appear in the fossil record until the Triassic, but must have evolved from one of these as-yet-unidentified anapsid groups. Generally, however, the anapsids never seem to have been as abundant or diverse as the synapsids or diapsids.

The diapsids, you may recall, are characterized by two postorbital fenestrae in their skulls. Primitive stem diapsids are known from the latest Pennsylvanian, but by the Late Permian, there is a recognizable separation of this group into two lineages: the lepidosauromorphs and the archosauromorphs. The lepidosauromorphs include the lizards, some of which were present during the latest Permian; the lepidosauromorph lineage would later give rise also to the snakes, sphenodontids (the New Zealand tuatara being the sole survivor), and Mesozoic marine reptiles (ichthyosaurs, plesiosaurs, mosasaurs). The archosauromorph branch is only known in the Permian from stem-group members, but is the stock that ultimately spawned crocodylians, dinosaurs, pterosaurs, and birds.

Neither the anapsids nor the diapsids attained the diversity and abundance during the Permian that the synapsids did, however. The synapsids sport a single postorbital fenestra and are the group from which mammals eventually evolved. In the Early Permian, the most successful group of synapsids was the pelycosaurs. Almost everyone will be familiar with the fin-backed pelycosaur *Dimetrodon*, frequently (though erroneously) called a 'dinosaur.' It was, instead, a representative of the sphenacodont pelycosaurs, the first group of large, terrestrial, carnivorous reptiles. The 'sail,' formed by elongated neural spines on the vertebrae, is thought to have been a thermoregulatory device; its size scales proportionately to the body weight of the animal. Calculations show that a 200 kg *Dimetrodon* could warm up from 26°C to 32°C in less than half the time with such a sail turned towards the sun as could a sail-less version. This would give a distinct advantage to a predator who could become active hours sooner in the morning than its prey, and is a strong indicator of a selective advantage for body temperature control. It indicates, however, that *Dimetrodon* did not yet possess an endothermic (warm-blooded) metabolism. Not all pelycosaurs had sails, but these did also appear in some representatives of the edaphosaurid lineage, such as the (probably herbivorous) *Edaphosaurus*—a likely example of predator-prey coevolution. The heyday of the pelycosaurs was the Early Permian, where they comprised 70% of the reptile genera and are known exclusively from equatorial areas, especially Texas and Oklahoma. By the Late Permian, pelycosaurs had died out, having been replaced by more advanced reptiles: the therapsids.

Fin backed pelycosaur *Dimetrodon*. Natural History Museum of Los Angeles County, Los Angeles, California.

The therapsids were a step further along the line toward becoming mammals than the pelycosaurs had been. Their dentition was frequently differentiated into incisors, canines, and cheek teeth—an indication of a better food-processing ability than the gulp-and-swallow method of the pelycosaurs. Chewing of food in the mouth exposes more surface area to digestive enzymes and increases digestive efficiency. This is important in an animal with high energy requirements, and probably indicates that some therapsids were well on their way to becoming warm-blooded. Another indication of a higher metabolism in some Late Permian therapsids is the development of a rudimentary secondary palate; this bony shelf separates the mouth from the nasal passages, making it possible for the animal to eat and breathe simultaneously—a novel feature. It has even been suggested that some therapsids may have had fur, although there is no fossil evidence for this (then again, fur rarely fossilizes even in much more recent mammals). By the Late Permian, the therapsids had replaced the pelycosaurs entirely as the dominant "mammal-like reptiles" (or "protomammals"). The therapsids also were posturally more advanced than their forebears, assuming more of an upright gait than the sprawling posture exhibited by their pelycosaurian ancestors. Movement of the limbs in a vertical plane is both energetically more efficient than sprawling locomotion, and allows higher speeds to be maintained.

We have mentioned in passing that some synapsids were vegetarians; the Permian saw the proliferation of herbivory amongst various groups of tetrapods—a very important development. Prior to the Permian, the main point of entry of plant energy into the food chain had been via detritivorous and herbivorous arthropods. With the evolution of vertebrate herbivory, a larger share of this plant energy was available to high-level consumers, which cut out the middlemen, so to speak. The proportion of herbivores dramatically increased in the Late Permian. By this time, then, ecosystems had finally reached a modern level of complexity.

The End-Permian Extinctions

The Permian ended with the greatest mass extinction known in the history of life on Earth. Although one may frequently read that "life on the planet was nearly wiped out," this is an overly dramatic statement. It is true that more than 90% of shallow-water marine invertebrate species disappeared forever, but if we look at the family level, the figure is closer to 70%. Equatorial species were hardest hit. The extinctions were less devastating to terrestrial faunas, which lost some 50-70% of

Transitional amphibian-reptile *Seymouria*, Cutler Formation, New Mexico. Denver Museum of Nature & Science, Denver, Colorado.

their taxa, depending on which measure is used. Plants suffered a slight decrease in diversity, but nothing monumental. Not to say that this wasn't a big deal—it obviously was a great catastrophe—but life on Earth came nowhere near to being exterminated altogether. Let's look at how some different groups fared.

Amphibian *Diplocaulus*, Vale Formation, Texas. Denver Museum of Nature & Science, Denver, Colorado.

Fusulinid foraminifera went extinct at the Permo-Triassic boundary, after having been in decline throughout most of the period. Other types of forams suffered substantially reduced diversity but pulled through and rediversified in the Triassic. Sponges were pretty much unaffected, as were conodonts. Tabulate and rugose corals all went extinct, but an unknown lineage of corals survived to produce the Triassic radiation of modern hexacorals. Bryozoans suffered a substantial reduction in diversity throughout the Late Permian, with only 20% of genera surviving the mass extinction. They persist to this day but have never regained their former abundance. Brachiopods fared in a similar manner; they were declining in diversity for some time prior to the boundary, but were nearly wiped out then. Ten genera pulled through and managed a Triassic recovery, but they never again were the dominant animals in their econiches, losing ground to the bivalve molluscs (clams), which suffered a less severe bottleneck. The gastropods (snails) are a 'Lazarus' group (so named as a Biblical reference to rising from the dead): they disappear from the fossil record entirely at the end of the Permian, only to reappear later in the Triassic (emphasizing the incompleteness of the fossil record). Nautiloids pulled through with a modest reduction in diversity, while ammonoids were hard hit. The few remaining trilobites and all of the graptolites went extinct (although there is a report of a recently-described (1993) animal that may possibly be a living graptolite). Sea urchins survived unscathed. The blastoids actually went extinct during the Late Permian several million years before the end of the period. Crinoids were almost exterminated, with the single genus *Isocrinus* surviving as a Lazarus taxon to effect a post-Triassic recovery. Fish survived virtually unscathed, with the loss of only a few archaic armored species. Stromatolites actually increased in abundance across the Permo-Triassic boundary, perhaps in response to the sudden dearth of marine grazers.

We can see from this that shallow marine invertebrates suffered a great crisis (reefs were practically wiped out and took up to 10 million years to recover), but that extinctions were distributed very inequally amongst different groups. Terrestrial vertebrates fared much better than most marine life. The pelycosaurs had already gone extinct millions of years earlier, about three-quarters of the way through the Permian. Amphibian diversity had been declining throughout the period, but this can be attributed to habitat reduction due to increasing aridity. Many groups of therapsids died out, but the cynodonts, which appeared on the scene in the latest Permian, survived into the Triassic where they gave rise to the true mammals. The archosauromorphs, which had been small, insignificant, and rare, passed the Permo-Triassic filter and went on to create the Mesozoic dynasty of dinosaurs. Plants, rather than being blindsided by the mass extinction event, underwent a gradual transition over a period of about 25 million years that saw the demise of the remaining large clubmoss and horsetail trees and the increasing abundance of cycads, ginkgos, and conifers.

What could have caused such widespread and devastating extinctions? By the mid-Permian, glaciation on Gondwana had ceased, so is unlikely to have been a factor. It has been suggested that a loss of continental shelf habitat resulting from the formation of Pangea increased competition and stress on shallow marine organisms; while this is certainly plausible, it doesn't account for the timing of the extinctions, which were largely abrupt and sudden at the end of the Permian. Although an asteroid impact has been proposed, the evidence for this is tenuous, and the suggestion that an impact could set off massive volcanism has been refuted by recent computer analysis. Volcanism may well have played a major role, however—the end of the Permian precisely (as well as current radiometric dating can determine) coincides with the most massive eruption of flood basalts (lava flows) ever known: the Siberian Traps, which contain enough material (an estimated 2 - 3 million km³) to cover the entire surface of the Earth with a layer 16 km (10 miles) thick. The best information to date indicates that this tremendous episode of volcanism took place within a timespan of only a million years or so.

How would volcanic eruptions cause extinctions, assuming you weren't the animal unlucky enough to be standing in the path of the lava? The Siberian Traps basalts are interleaved with numerous ash beds, indicating that a tremendous amount of dust erupted at the same time relative to what is typical of most flood basalts. Volcanic ash and gases—particularly sulfur dioxide and carbon dioxide—have known climate-altering effects, as well as being immediately poisonous to animals in the vicinity of the eruption. The biggest eruption of the 20th century, Mt. Pinatubo, was tiny by comparison but lowered global temperatures by half a degree C (1°F) the following year. This may not sound like much, but it was enough to make the summer of 1992 unusually cold, with bitter rain and wind spoiling our family vacation to Lake Michigan in July. The largest

Permian faceoff:

Dimetrodon limbatus &
Eryops megacephalus
(casts), Admiral Formation,
Archer Co., Texas. Denver Museum of
Nature & Science, Denver, Colorado.

eruption in historical memory was in Iceland in 1783-84, and involved an estimated 12 km^3 of volcanic material. Eyewitness accounts attest to the devastation of crops and forests over most of the island, and records indicate a worldwide drop in temperature of about 1°C (2°F). The effects on climate of the continuous eruptions necessary to produce the Siberian Traps are truly incomprehensible. Initially, SO_2 and ash particles block sunlight, leading to a global cooling, but later, as these clear from the atmosphere, the effect of CO_2 takes over and induces global warming. But is a modest warming, in and of itself, sufficient to account for the magnitude of these extinctions?

Possibly, and there is corroborating evidence. Loss of the cold-adapted, high-latitude Gondwanan *Glossopteris* flora supports global warming. Also, remember how part of the definition of the Permo-Triassic boundary is 'the termination of a major negative carbon isotope excursion'? What does that mean? Carbon comes in various naturally-occurring forms (isotopes), two of which are ^{13}C and ^{12}C. (A third, ^{14}C, is radioactive and used in the familiar 'radiocarbon dating'.) Carbon isotopes are recorded in the calcium carbonate shells of marine organisms, in inorganic carbonate sediments, and even in the mineralized portions of animal bones; the isotopic ratio preserved in these fossils reflects the ratio of available carbon isotopes in the environment at the time. Methane (CH_4), which occurs on the seafloor as methane hydrate, a solid, ice-like substance, is isotopically very 'light,' meaning it has a low ratio of ^{13}C to ^{12}C compared to the environment as a whole. Current estimates place the amount of methane hydrate in continental slope sediments at between 10 and 20 x 10^{15} kg (quadrillion kilograms), a tremendous amount. A similar amount may have existed on the Permian seafloor, and methane hydrate is very temperature-sensitive with regard to its stability. A small rise in temperature can cause the discharge of massive amounts of this buried methane hydrate into a gas, which rises and enters the atmosphere. The entry of a large amount of isotopically light methane into the ocean and atmosphere—between 10% and 25% of what is contained in the oceans today—could cause the carbon isotope excursion seen in the sedimentary record at the end of the Permian. Methane is a far more powerful greenhouse gas than carbon dioxide. If the initial atmospheric warming caused by volcanically-introduced CO_2 were to warm the oceans sufficiently to cause the dissociation of methane hydrates, a positive-feedback loop would be set in motion that led to runaway global warming. In fact, oxygen isotope studies indicate a rapid warming of about 6°C (11°F) in equatorial waters coincident with the Permo-Triassic boundary.

Cotylosaur *Diadectes phaseolinus,* Godlin Creek, Archer Co., Texas. American Museum of Natural History, New York.

But how does this affect marine life? Don't organisms like corals like warm waters? In moderation, yes—but as we unfortunately hear about all too frequently these days, too much warmth can cause corals to "bleach" (expel their symbiotic algae, called zooxanthellae)—and subsequently die (one result of human-induced 'global warming.') And there's more to the story. There is further evidence that there was widespread marine hypoxia and anoxia (low or no oxygen) concurrent with the Permo-Triassic extinction event. How do we know? Radiolarian-bearing cherts from British Columbia and elsewhere record the widespread presence of reddish hematite (an oxidized mineral of iron) in sediments until the very latest Permian, when suddenly they turn black, contain pyrite instead of hematite, and become almost devoid of radiolarians—all indications of anoxia—for an interval extending into the earliest Triassic, after which they return to normal. The solubility of oxygen is reduced in warm water, so low oxygen and high water temperatures go hand-in-hand. If the warming is accompanied by a lessening of the equator-to-pole temperature gradient, oceanic circulation patterns are weakened, which also promotes anoxia. And a lack of oxygen spells doom for marine animals; they simply suffocate.

With an event that happened a quarter of a billion years ago, it is understatement to say that unravelling the complex web of factors responsible for the extinction is difficult. The scenario outlined above is the one that I find the most convincing, but alternate theories abound. Volcanically-induced global warming—leading to the release of seafloor methane hydrates and consequent anoxia—would account for why marine creatures were devastated preferentially over terrestrial animals, and for why, in general, terrestrial vegetation seems to have been little affected. But in the absence of a time machine, there remains plenty of room for research and speculation.

Where Permian Rocks Are Found

Permian outcrops are entirely lacking east of the Mississippi River, and Permian rocks in general are less well-exposed in North America than those from other periods of the Paleozoic. The redbeds of west Texas and southeast New Mexico have already been mentioned, and large expanses of Permian sediments also outcrop in north-central Texas, across half of Oklahoma, and into east-central Kansas. The Kaibab Limestone of Permian age forms the rim of the Grand Canyon and is well-exposed across northern Arizona. Elsewhere Permian outcrops in North America are few and far between, and they are largely nonexistent outside of the States.

A creative way of restoring the pelycosaur *Edaphosaurus* when only the vertebral column was found. American Museum of Natural History, New York.

Where Permian Fossils Are Found

There are a few localities that have produced spectacular Permian fossils, however. The Gaptank Formation of the Glass Mountains of southwest Texas is a goldmine for beautifully-preserved ornate brachiopods, as well as algae, sponges, bryozoans, crinoids, conodonts, corals, gastropods, nautiloids, ammonoids, and ostracods. The Admiral and Vale formations of Texas, the Arroyo Formation of Oklahoma, and the Cutler Formation of New Mexico are some of the best places in the world to find Permian tetrapod bones. The Wellington Formation of Kansas and Oklahoma has produced a spectacular insect fauna. Arizona's Kaibab Limestone contains brachiopods, shark's teeth, crinoids, horn corals, sponges, and other marine invertebrates. The underlying Hermit Shale, Supai Formation and Coconino Sandstone are also Permian in age, and record arid terrestrial sediments. The Hermit Shale yields a wealth of fossils of xeric seed plants, ferns, and tetrapod tracks; footprints can also sometimes be found in the Supai and the Coconino. Further north, exposures of the Phosphoria Formation in Idaho yield sponge spicules, horn corals, bryozoans, brachiopods, ammonites, and clams.

Where Can Permian Fossils Be Seen in the Field?

Given the dearth of exposed Permian sediments, travellers are in fairly good shape when it comes to finding publicly-preserved outcrops, and their occurrence in arid and semiarid climates makes the fossils relatively easy to find. Hueco Tanks State Historic Park near El Paso, Texas contains outcrops of the Magdalena Formation and Hueco Limestone, which are chock-full of beautiful bryozoans, brachiopods, clams, crinoids, corals, snails, nautiloids, and ammonites. The Capitan Limestone, and the El Capitan monolith itself, are part of the Capitan reef complex and can be seen along the Permian Reef Geology Trail in Guadalupe Mountains National Park. This preserve, located in the Guadalupe Mountains of west Texas, is a good place to see Permian clams, echinoids, horn corals, the rare trilobite, and other remnants of a reef fauna.

We've previously mentioned caves as sometimes good places to find fossils. Arguably the most famous cave in the United States, and certainly one of the largest and most beautiful, is Carlsbad Caverns, located near the town of Carlsbad in southeast New Mexico. The caverns are carved out of the Permian Carlsbad Limestone; take time on your self-guided tour to look in the cave walls (if you can find a spot not covered with flowstone!) for marine invertebrate fossils, including sponges, brachiopods, bryozoans,

Permian driftwood, provenance unknown. New Mexico Museum of Natural History, Albuquerque.

nautiloids, crinoids, and clams. You may also have some luck spying similar fossils along some of the hiking trails in the park.

As noted previously, the rim of the Grand Canyon is formed by the Permian Kaibab Limestone. Ranger-led "fossil walks" leave twice a day from Bright Angel Lodge on the South Rim, and are an excellent way to become acquainted with the fossils of the area. You can see a variety of brachiopods, corals, molluscs, and crinoids in the Kaibab, as well as various reptile tracks in the Coconino Sandstone and good plant fossils in the underlying Hermit Shale.

Not too far from the Grand Canyon, a few miles east of Flagstaff, Arizona, is Walnut Canyon National Monument. Although most visitors are drawn to Walnut Canyon because of the cliff dwellings it harbors, it too exposes the Coconino Sandstone and Kaibab Limestone, where you can see Permian fossils similar to those noted for the Grand Canyon.

Glen Canyon National Recreation Area, which straddles the Utah-Arizona border, is a bit more remote, but still worth a trip. There, exposures of the Cedar Mesa Formation contain abundant vertebrate tracks.

When so many econiches are vacated as happened in the end-Permian mass extinction, the future belongs to a few survivors. Inevitably, these radiate to occupy vacant habitats and the new landscape looks little like the old. The fossil record indicates that it took some 5 - 10 million years for the biosphere to recover from the terminal Permian event. It was the close of one era and the dawn of the next. The new would belong to those little, insignificant diapsid reptiles that up until now were only bit players: the archosauromorphs and their most successful spawn, the dinosaurs.

The Mesozoic Era

Monsters ruled the land and sea during the Mesozoic. In this snapshot of the Kansas ocean 70 million years ago, the mosasaur *Tylosaurus* and the terror fish *Xiphactinus* vie for a hapless ammonite. (Will he get away?)

Although we have covered more than half of the Phanerozoic Eon—the span for which we have a good fossil record of animals and plants—we are only now embarking on our journey through the "time of middle life," or Mesozoic Era. There are two main reasons for this skewed coverage. As life becomes more "evolved," there is more to interest the general reader (I'll bet a nickel there are more of you captivated by *Tyrannosaurus rex* than by an earthworm). But fundamentalist-religious and anthropocentric points of view aside, it's rather difficult to say that life today is really more complex than it was during the Cambrian; there is no "ladder of progress" as envisioned by Victorian naturalists. Readers of the late S. J. Gould's writings are familiar with his arguments for the central role of contingency (luck) in the course of evolution, a concept which I embrace. Roll the dice again, and the history of life would most likely come up altogether different.

But there is another, more important reason for increased attention to the recent: the fossil record gets better the closer we get to now. This is simply the result of there having been less time for those fossils to have been eroded away. Some sediments are deposited during every geologic age, although the amount of them laid down varies radically depending on the environments present (much in shallow seaways, little in dry areas). Given a long time, most of those sediments will be removed by erosion. When a shorter span has passed, Mother Nature has had less time to do her destructive work. More sedimentary rock—more fossils—more research—more is known.

By the dawn of the Mesozoic Era, terrestrial ecosystems had attained a level of complexity comparable to that of modern ones: stratified forests, and a food chain supported by plants, eaten by vertebrate and invertebrate herbivores, which in turn were consumed by large and small carnivores. But a vastly different set of players occupied the stage. The Mesozoic began without any mammals; it saw the rise and fall of the dinosaurs (although they live on among us today as birds); and there were initially no flowering plants. By its close, arguably all the major developments in evolutionary history had occurred, and a time traveller would not feel too out of place either on the land or in the seas. Major extinctions marked both its beginning and its end—and in a very literal sense, the Mesozoic went out with a "bang." But we are getting ahead of ourselves. Let's start with life's recovery from the end-Permian catastrophe and go from there.

Chapter Nine
The Triassic

The western waves lap at the foot of a small mountain range in central Nevada as the "Age of Reptiles" dawns. An ichthyosaur, *Shonisaurus*, cruises near the surface of the water, paying little heed to the fish and ammonites swimming nearby.

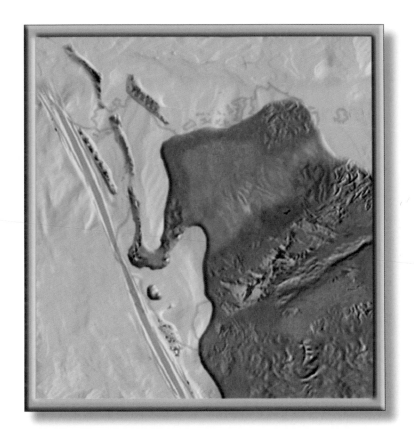

Triassic North America

The Mesozoic Era dawns with the onset of the Triassic Period. The "Era of Middle Life" would see the replacement of the archaic Paleozoic fauna and flora with ecosystems that were considerably more modern in aspect, yet far enough removed from our time that they are still unquestionably exotic and fascinating. In our imaginations the Mesozoic is inextricably tied to the dinosaurian dynasty; yet the appearance of another, smaller, but no less important group—perhaps one even moreso from the human perspective—also took place about the same time, in the Late Triassic: the mammals.

Definition and Nomenclature

The name 'Trias,' still used in its original form by many European geologists, was given to the first period of the Mesozoic by Friedrich August von Alberti in 1834. 'Trias' is the Latin word for 'three,' and von Alberti chose it to emphasize the tripartite nature of the sedimentary record in southern Germany during this timeframe: the Bunter, or Lower Triassic, which contains brown sandstones and 'redbeds'; the Middle Triassic Muschelkalk, a bivalve-rich, predominantly marine limestone series; and the Keuper, or Upper Triassic, consisting of terrestrial rift-basin deposits containing abundant dinosaurs and other vertebrate fossils. Although the name Triassic is used everywhere, this threefold division is only a local phenomenon.

The Permo-Triassic boundary, or start of the Triassic, is defined as the lowest occurrence of the conodont *Hindeodus parvus* and the termination of a major negative carbon isotope excursion at the base of 'bed 27c,' Meishan, Zhejiang, China. This event is currently dated at 251 mya. The end of the Triassic, or Triassic-Jurassic boundary, has yet to be tied to a global stratotype section, but has a working definition of "near the lowest occurrence of the smooth *Psiloceras planorbis* ammonite group," with a U-Pb radiometric date of 200 mya. The Triassic may thus be considered to have lasted some 51 million years.

Paleogeography and Paleoclimate

The supercontinent of Pangea encompassed all the world's landmasses when the Triassic dawned. It persisted throughout the period, although by the Late Triassic, the rifts were forming that would eventually rend it asunder, creating the familiar seven continents of today. Pangea stretched essentially from pole to pole, and the other side of the world was occupied by the vast Panthalassic Ocean. In the east, the equatorial Tethys Seaway separated Asia from East Gondwana, but in the west, North America was still firmly sutured to South America and Africa; you could have easily trekked from the Arctic to the Antarctic without getting your feet wet. The whole of Pangea was drifting slowly northward, such that by this time, most of North America was in the northern hemisphere. The equator ran approximately from El Paso, Texas through the Florida panhandle.

Petrified wood *Araucarioxylon arizonicum*, Chinle Formation, Holbrook, Arizona. Author's collection.

You will recall that the close of the Permian brought with it a dramatic global warming (possibly due to release of seafloor methane stores), and this 'hothouse' climate persisted throughout the Triassic; this was, in fact, one of the hottest times in Earth's history. Because of the rising of warm, solar-heated air at the equator, condensation of moisture occurs as this air cools, and there is a tropical belt from approximately 10°N to 10°S that typically receives a lot of rainfall. After voiding itself of moisture, this air mass bifurcates and heads toward the midlatitudes, where, now dry, it descends at around 30° north and south latitude, creating persistent zones of high pressure and aridity. Lands in these dry belts are frequently deserts unless they have access to reliable maritime moisture.

These were the conditions that existed in North America during the Triassic. To start, much of the United States and Mexico was in the wet equatorial tropics, but the rains were unreliable in the interior regions of the continent, far from the sea. The supercontinent also generated strong monsoons, such that much of the rainfall these areas received was probably seasonal. Canada, on the other hand, firmly straddled the northern arid belt, and played host to large expanses of desert. The climate was hot and dry enough for salt deposits to form as far north as British

Columbia, and there were large expanses of sand dunes in Nova Scotia. Throughout the Triassic, Pangea continued to drift slowly northward until the continental U.S. found itself in the dry, high-pressure zone. So, in addition to the widespread aridity found in the interior regions of Pangea because of their distance from the sea, from the perspective of the central areas of our continent, the climate became increasingly dry as the Triassic progressed. This is evidenced by the newfound presence of nodules and crusts of caliche in paleosols (ancient soils) from the eastern U.S. and by the increasing abundance of xeric (dry-adapted) seed plants in the flora. Marine rocks from the Triassic are rare because sea levels were generally low; no geological time period is represented by fewer marine deposits in North America than the Triassic. This may be related to the general lack of tectonic activity and reduced spreading at the midocean ridges: hot rock occupies more volume than cold rock, thus when seafloor spreading rates are high, the oceans spill onto the low-lying areas of the continents, and when tectonic activity is low, the seas retreat. Continental sediments from this timeframe are typically red sandstones and evaporites formed under arid conditions. The few North American Triassic marine deposits are all found in the West, where parts of Nevada, Utah, Idaho, British Columbia, and Alaska formed a back-arc basin behind a chain of mountainous, volcanic island terranes (the McCloud Arc and Sonoma Mountains). Subduction of the Pacific plate beneath the western margin of North America caused the accretion of many small fragments of land to the Cordilleran margin, and modest mountain-building occurred from Alaska to southern Mexico.

Life in the Triassic

Earliest Triassic faunas were extremely depauperate in the wake of the end-Permian extinctions. It took some 5 - 10 million years for reefs to recover; they were formed now by the modern hexacorals, which built small patch reefs over the moribund remains of earlier sponge reefs. It would be some time before reef faunas approached the diversity and abundance they had enjoyed during the Devonian, or even the modest success they had attained by the Late Permian. Nevertheless, they were on the road to recovery, while undergoing extensive faunal remodelling as the few survivors of the end-Permian event radiated to fill vacant econiches. Brachiopods, which had dominated benthic (seafloor) habitats during the mid-to-late Paleozoic, were largely replaced by bivalved molluscs; although a few brachiopods survive to this day, they are relegated to the role of bit players. In some places bivalves (clams) were so numerous that the rock is made up of little other than their fossilized

Coelacanth (lobe-finned fish) *Diplurus newarki,* Triassic rift valley lakes, Princeton, New Jersey. American Museum of Natural History, New York.

shells. Ammonoids rebounded, and types with more complex sutures evolved; belemnites, squidlike relatives with an internal shell, first appeared. Gastropods (snails) reappeared after a seeming gap—they likely survived the Permo-Triassic filter as small, relict populations whose fossils have yet to be found—and effected a stunning recovery. Crinoids hung on by a thread with a single surviving genus, but their rediversification was essentially post-Triassic. Bryozoans and echinoids (sea urchins) were present but uncommon. Fishes, largely unaffected by the Permo-Triassic event, were becoming increasingly modern as the new ray-finned fish enjoyed accelerating success at the expense of more archaic types, with the exception of sharks, which continued to reign as top predators.

By the Triassic, though, those sharks had competition, not from other fish, but from a secondarily aquatic group of diapsid reptiles, the ichthyosaurs ('fish lizards'). Some of the first ichthyosaurs are known from the Early Triassic of Canada. The streamlined, dolphin-like shape and sharp teeth of the ichthyosaurs indicate that they were fast, formidable predators. Early Triassic ichthyosaurs were already fully adapted to a marine existence, and although air-breathers, they gave birth to live young in the water—a fact demonstrated by at least one fossil of an ichthyosaur mother who died in childbirth, with the young half in and half out of the birth canal. Ichthyosaurs increased in size throughout the Triassic, with the Late Triassic genus *Shonisaurus*—known from more than a dozen skeletons from the Luning Formation at Berlin-Ichthyosaur State Park in Nevada—attaining a length of some 15 meters (50 feet), the largest member of this group known.

Dinosaur *Coelophysis bauri* (cast), Chinle Formation, Ghost Ranch, New Mexico. Natural History Museum of Los Angeles County, Los Angeles, California.

Late in the Triassic, another quintessential group of Mesozoic marine reptiles appeared on the scene—the plesiosaurs. They are thought to have evolved from the nothosaurs, a long-necked group of diapsids which was somewhat, though not fully, adapted to an aquatic existence. The largest plesiosaurs rivalled the largest ichthyosaurs in size, being up to 15 meters (50 ft) long, comparable to the size of a modern killer whale. Most people will picture a 'Loch Ness Monster' type of creature when bringing a plesiosaur to mind, but the small-headed, long-necked forms are only typical of the plesiosaurid branch of the group. Another group, the pliosaurs, was characterized by shorter necks and larger heads than the plesiosaurids. It is not known whether plesiosaurs gave live birth in the ocean or whether they continued to come ashore to lay eggs. Their strong flippers may have been sufficient for rudimentary land propulsion, but if so, they would doubtless have been awkward ashore as are modern seals. The teeth of both ichthyosaurs and plesiosaurs indicate that they dined primarily on fish.

And for the first time, the air belonged to more than just the flying insects. The earliest pterosaurs appeared near the end of the Triassic—the rhamphorhynchoids, a group characterized by long necks and tails and short faces. These diapsids were the first group of vertebrates to evolve active flight, and it is likely that they had a high metabolic rate to support such strenuous activity. By the time they appear in the fossil record, they are fully adapted to an aerial lifestyle. Flight has evolved separately three times in vertebrate history: first, in pterosaurs in the Late Triassic; next, in birds in the Late Jurassic; and finally, in bats during the Early Tertiary. Each time, nature found a different solution to designing the wing. Pterosaurs supported the wing membrane with an elongated fourth finger; birds, with the bones of the arm and hand; and bats, with multiple fingers fanning out from the apex of the wrist. Pterosaur fossils are most frequently found in shallow marine deposits, indicating that many species were probably shore-dwellers that fed on schools of fish. There are two competing hypotheses regarding how they developed flight—the "ground up" and the "trees down" scenarios. In the first, running pterosaurs learned to generate lift, and later to fly; in the second, flight began as gliding behavior. The jury is still out on which is the correct model.

Terrestrial floras showed a continuation of the trend towards the dominance of dry-adapted seed plants (gymnosperms) that had begun in the Permian, at least in the northern hemisphere. In North America, conifers were the primary large forest trees throughout the Triassic, complemented by ginkgos, cycads, and a new group, the bennettitaleans or cycadeoids. Cycads and cycadeoids, while similar in morphology (many looked somewhat like overgrown pineapples with palmlike leaves), are thought to have evolved separately from seed fern ancestors. "Cycadophytes" is a term that is sometimes used to refer to cycads and cycadeoids together, but this is an unnatural grouping based on form rather than ancestry.

In the wetter areas, such as montane forests and riparian (river- and lakeshore) habitats, a few fairly large horsetails and the single clubmoss tree *Sigillaria* made their last stand. The horsetail *Neocalamites* attained the height of a telephone pole, much smaller than its Pennsylvanian 'coal swamp' relatives but far larger than any modern type nonetheless. Tree ferns also favored the moister areas, along with herbaceous ferns. In the high northern and southern latitudes—in places such as Greenland—coals formed, indicating that these areas enjoyed a reliably cool, moist climate.

Petrified wood *Araucarioxylon arizonicum*, Chinle Formation, Holbrook, Arizona. Author's collection.

The earliest Triassic saw the almost complete dominance of the land by herds of the therapsid (mammal-like reptile) *Lystrosaurus*. Most of the known fossils of these animals come from Gondwana, from a few very rich sites. The sedimentary record of the northern continents from this timeframe is very poor, so it is not known if these animals ruled in the north as well. Two types of therapsids, cynodonts and dicynodonts, had emerged in the latest Permian to become the most successful members of the group. 'Cynodont' means 'dog tooth' and is a reference to the increasingly specialized dentition of these synapsids. Instead of the generalized homodont ('same teeth') condition seen in most Permian pelycosaurs, advanced therapsids had heterodont ('different teeth') dentitions: their teeth were differentiated into anterior incisors, larger canines, and cheek teeth which were specialized for their particular diet. The cynodonts also developed a secondary palate, which, you will remember, is a bony shelf that separates the mouth from the nasal passages, allowing the animal to eat and breathe at the same time—important in an animal with high food requirements. Such high energy intake was probably necessitated by an endothermic (warm-blooded) metabolism, which was evolving during the Late Permian and Early Triassic. Contrary to popular belief—which is based on the discrete difference between the warm-blooded mammals and birds on the one hand, and the cold-blooded reptiles on the other, which inhabit the modern world—endothermy is not an "all or nothing" condition. Although modern reptiles lack metabolic control of their body temperature, many of them maintain a remarkably constant internal temperature by behavioral means (moving from sun to shade or vice versa). Also, although we, as humans, tend to think that body temperature "should" be close to 37°C (98.6°F), this is far from universal. Birds frequently maintain an internal temperature several degrees higher than we do, and the duck-billed platypus, a primitive, relict egg-laying mammal that lives in Australia, only maintains about 30°C (86°F). Thus, there are varying degrees of endothermy, and it is likely that such metabolic temperature control was achieved in stages in the cynodonts. That they did so is supported by fossil evidence of an upright posture, differentiated dentition, and secondary palate; additionally, fossils of some Triassic cynodonts show pits along the snout that in modern mammals contain nerves that supply the vibrissae, or whiskers, with sensation. The argument goes that if they had whiskers, they had fur, and if they had fur, they were trying to keep warm. However, the degree of metabolic temperature control that these advanced cynodonts maintained is open to speculation. Another feature we tend to associate with mammals, their large brains, had yet to evolve, however: preserved braincases of cynodonts show no bigger brains than would be expected in reptiles of similar size. Thus, they probably weren't overly intelligent.

These therapsids shared the Triassic world with the thecodont archosaurs and the earliest crocodylians, as well as the first known turtles. Crocodiles are essentially unchanged since their Triassic appearance and give us a good glimpse into an ancient fauna. There were several main groups of thecodonts: the herbivorous, armored aetosaurs, which looked something like giant, partially-flattened armadillos; the phytosaurs, semiaquatic reptiles that in body form and lifestyle greatly resembled the crocodylians but were not closely related; and the rauisuchians, which were large predators with powerful hind limbs that may have made them facultatively bipedal (meaning they spent most of their time on all fours, but could rear up, and possibly run short distances, on two legs if they wanted to). The Chinle Formation of Arizona and the Dockum Formation of Texas have yielded fossils of entire Triassic ecosystems which can guide us in reconstructing the thecodont-dominated world. Aetosaurs peacefully grazed on low vegetation (ferns and cycads), phytosaurs were fish-eaters, and the rauisuchians were the terrors of the land, chasing down therapsids, aetosaurs, and anything else they could catch. Terrestrial thecodonts paralleled the therapsids in evolving a more upright gait than their ancestors, and hence increasingly efficient locomotion. There were even slender, upright "running crocodiles" that were terrestrial predators.

Temnospondyl amphibian *Buettneria* sp., Herring Ranch, Potter Co., Texas. American Museum of Natural History, New York.

By the Late Triassic, the cynodonts had spawned the true mammals, and the remaining therapsids were on the wane. The earliest unquestionable fossil mammal is from the Late Triassic of Texas. What makes a mammal? We have already discussed warm-bloodedness. Another defining feature is the presence of mammary or milk glands, for which the group is named. Embryological studies have shown that milk and sweat glands are very closely related developmentally; sweat glands again are a method of temperature control, albeit one of cooling rather than of keeping warm. It is thus logical to assume that the production of milk followed on the heels of the evolution of fur (which necessitated sweat glands). The platypus, in fact, secretes its milk from glands distributed among the hairs of the breast region which are not grouped into a teat—probably the ancestral condition. Concomitant with this is the change in the pattern of tooth replacement from a continuous one, as is seen in reptiles (and cynodonts), to the mammalian condition whereby most of the teeth are replaced only once, and the molars not at all; and the evolution of multi-rooted, rather than single-rooted, molars. These changes allow for more precise dental occlusion and better food-processing efficiency, but the delay in tooth eruption required to allow for only a set of 'baby' (or 'milk') teeth and an adult set necessitates that the young be initially nourished in a manner that requires no chewing; and, in mammals, this is mother's milk.

Late Triassic mammals were small—the size of a mouse or a shrew—and from the predominance of rods rather than cones in the retinas of primitive living mammals' eyes, it has been inferred that early mammals were probably nocturnal. Some may have lived in burrows, where their sensitive whiskers helped them navigate. The earliest mammals were brainier than their cynodont ancestors but less so than modern mammals of comparable size; the increase in intelligence has been suggested to be a response to the trials of navigation in the dark and in tunnels. Another late-appearing feature was live birth, which may not have arrived until the late Mesozoic. Taking the living monotremes as an example (the platypus and echidna of Australia and nearby islands), it is clear that it is possible to be a mammal and yet still lay eggs. Such was almost certainly the case among the earliest mammals.

Most of the fundamental differences between mammals and modern reptiles can be accounted for by the former's high metabolic rate. While this is energetically costly (a warm-blooded mammal typically requires about ten times the amount of food as a cold-blooded reptile of comparable size), the benefits are enormous. Mammals rely on oxidative metabolism to power their muscles over the long term, whereas reptiles such as lizards use anaerobic metabolism. What's the difference? Anaerobic metabolism doesn't require a sustained oxygen intake (or an efficient respiratory and circulatory system), but produces lactic acid as a by-product, which causes muscle fatigue. A lizard can sprint for short distances as fast as a mammal, but after a minute or two of this activity, he is exhausted and needs several hours to recover. Aerobic (oxidative) metabolism produces substantially more energy, and generates only oxygen and carbon dioxide as by-products, so the animal can sustain a high activity level for long periods of time without becoming fatigued. Coupling this ability with an upright, rather than a sprawling, posture—which makes locomotion more efficient—we can see why the early mammals were in a position to supplant "traditional" reptiles in their chosen econiches.

However, it was not yet time for the mammals to inherit the Earth. In the mid-to-late Triassic, some tough competition appeared—the first dinosaurs. Although the oldest unquestionable fossil dinosaur bones are from the Late Triassic, mid-Triassic footprints ascribed to dinosaurs indicate that they had arrived on the scene. Descended from archosauromorph ancestors, dinosaurs were the group which would achieve the greatest success among the diapsids—at least, if measured by the length of time they reigned as the dominant animals on the land. Several features define a dinosaur. The head of the femur (thigh bone) is inturned at close to a 90° angle, and the shaft of the femur is straight, or nearly so: this causes the leg to be borne upright, and move in a vertical (fore-aft) plane. By attaining an upright posture, inward horizontal force from the leg against the hip socket is reduced; in dinosaurs a perforation of the socket occurs due to this reduction in stress, although it

Adult & juvenile dinosaurs *Coelophysis bauri*, Chinle Formation, Ghost Ranch, New Mexico. Denver Museum of Nature & Science, Denver, Colorado.

may have been filled in with cartilage in life. At the same time that medial forces are lessened, the upper rim of the socket has to bear more weight, and thus it develops a reinforcing ridge.

Almost from the start (by around 230 mya), dinosaurs split into two important lineages: the ornithischians and the saurischians. These groups are differentiated based on the shape of the pelvis (hip bones). The saurischian ('lizard-hipped') dinosaurs have a triradiate (three-pronged) pelvis with an anteriorly-pointing pubic bone. The ornithischian ('bird-hipped') dinosaurs have a pubic bone that is rotated posteriorly to lie alongside the ischium. It is one of the ironies of nature and nomenclature that birds are descended not from the bird-hipped, but from the lizard-hipped, dinosaurs.

Most Triassic dinosaur fossils are poorly-preserved and not well known. There is sufficient evidence to indicate that saurischians, in the guise of small, meat-eating theropods and larger (up to 8 m/26 ft) vegetarian prosauropods, were present. Ornithischians are represented by the fabrosaurs and heterodontosaurs, two small, unspecialized forms. The only really well-known Triassic dinosaur is *Coelophysis*, a small carnivore whose articulated skeletons are found by the dozens at an amazing lagerstätte at Ghost Ranch, New Mexico. This concentration of bones in the redbeds of the Late Triassic Chinle Formation may be the remains of a herd of *Coelophysis* who were drowned in a flash flood, or alternatively, the mass mortality of animals that had congregated around a drying water hole. *Coelophysis* became the first "dinosaur astronaut" in January of 1998 when a perfect skull was taken aloft by the shuttle Endeavor and transferred to the Space Station Mir.

Cycadeoid leaf *Zamites powelli*, Chinle Formation, Utah. Denver Museum of Nature & Science, Denver, Colorado.

Once the dinosaurs had arrived on the scene, they supplanted the thecodonts and the therapsids, and relegated the mammals to skulking in the shadows. Much of their success has been attributed to their upright posture, and hence fast running ability; but we have seen that cynodonts and mammals possessed this also. Why did the dinosaurs triumph? The suggestion has been made that in the case of the prosauropods, their long necks and tall stature when standing bipedally allowed them to access previously-unexploited food resources high up in the trees. This certainly makes sense, but for that group alone. In what way were dinosaurs, in general, superior to therapsids and mammals? We must look to their contrasting physiologies and the climate of the time for a plausible answer.

All animals produce metabolic waste products, and must dispose of them. Synapsids—here, therapsids and mammals—produce urea, which must be excreted while dissolved in abundant water. Diapsids (dinosaurs, in the present discussion, but also including lizards, birds, and crocodylians) instead excrete uric acid, which is much more water-efficient. We have already seen that the climate of much of Pangea was hot and dry, at least seasonally. We should mention now that the earliest dinosaurs appear in those hot, dry regions. It may very well be that they outcompeted their thecodont relatives because of superior locomotor ability, and rose to dominance over the therapsids and mammals because they were better-adapted to a dry climate. It is always difficult to understand evolution in retrospect, but the above explanation is beguiling. By the latest Triassic, dinosaurs were the most common tetrapods in dry, non-equatorial basins, and they occupied nearly every terrestrial econiche.

The End-Triassic Extinction

The Triassic started after a mass extinction, and ended with another one. The latter was not so severe, and the only major marine group lost was the conodonts; nevertheless, some 40% of marine invertebrate genera were wiped out. Again, terrestrial ecosystems were not as hard hit as were those in the seas, with more species turnover than loss of major groups. As with the end-Permian event, all the usual culprits have been proposed (climate change, sea level fluctuations, fragmentation of Pangea, asteroid impacts, volcanism). A few years ago some scientists thought they had zeroed in on the asteroid that caused the extinctions *and* the crater it formed: a ring-like, lake-filled, 70-km-diameter structure at Manicouagan, Québec, that seemed to be just about the right age. However, more recent and precise dating of both the Manicouagan crater and the Triassic-Jurassic boundary appears to have nixed that scenario—the crater is some 12 to 14 million years too old to coincide with the extinctions. So we are back to searching for a terrestrial perpetrator. The leading candidate at this point appears to be another flood basalt episode and its concomitant environmental disruption. The Central Atlantic Magmatic Province (CAMP) was emplaced during the waning years of the Triassic and the beginning of the Jurassic as Pangea rifted apart along North America's eastern edge. Estimates of the amount of volcanic material initially involved vary widely, but usually are around 1 to 2 million km^3 (although some contend that it was a larger flood basalt episode than the Siberian Traps). Even the low end of this range would be a tremendous event, and could bring with it all the environmental effects we discussed in the last chapter in connection with end-Permian volcanism. CAMP basalts are found today on four continents (North and South America, Africa, and Europe), although most of the North American share is buried under sediments on the continental shelf. Dike swarms associated with the East Coast rift valleys are more visible evidence of this episode.

Where Triassic Rocks Are Found

Triassic sediments are scarce in North America, and marine units particularly so. In the east, sandstones and shales laid down in the rift valleys that formed prior to the breakup of Pangea are exposed in narrow bands from Massachusetts to North Carolina. At times, these rift valleys held lakes similar to those seen in east Africa today—lakes Malawi and Tanganyika are examples—and into which sediments of the Newark Supergroup and correlative formations were laid down.

Other than that, all of the Triassic sediments in North America are west of the Mississippi River. Exposures are absent from the midcontinent, but fair-sized outcrops are found in west Texas and the Four Corners region of Colorado, New Mexico, Arizona, and Utah. The most famous of these is the Chinle Formation, familiar to many travellers as the home of Petrified Forest National Park. Smaller Triassic outcrops can be found in Nevada, Idaho, the central mountains of British Columbia and the Yukon, on the North Slope of Alaska, and in the Canadian Arctic isles.

Where Triassic Fossils Are Found

The various formations of the Newark Supergroup of the Atlantic States harbor the fossils of ferns, horsetails, conifers, cycads, and freshwater fish. This is also one of the richest rock units in the world for dinosaur tracks, particularly in the Lockatong and Passaic formations. Reptile bones and archosaur teeth have been reported from a small outcrop of the Carr's Brook Formation in Nova Scotia.

The Chinle Formation, mentioned above, produces phytosaurs, aetosaurs, rauisuchians, metoposaurs (large amphibians), lungfishes, sharks, coelacanths, cycads, horsetails, cycadeoids, and other plant fossils in addition to the conifer logs for which the Petrified Forest is named. Most of the petrified wood found in Arizona, frequently offered for sale at roadside stands, comes from the Chinle. The Dockum Formation of Texas has produced a similar biota. The Chinle in New Mexico contains the Ghost Ranch Quarry, the only place where Triassic dinosaur bones are abundant in North America. In Utah only plants and tracks have been found in this formation. Underlying the Chinle is the Moenkopi Formation, which is rich in the footprints of thecodonts and amphibians as well as their bones, and contains tracks which are the oldest evidence of dinosaurs in North America. Phytosaur remains, especially teeth, have been found in the Bull Canyon Formation of New Mexico.

Phytosaur *Phytosaurus megalodon*, Dockum Formation, Texas. University of Michigan Exhibit Museum of Natural History, Ann Arbor.

Ammonoids can be found in the Thaynes Formation of Utah, Nevada, and Idaho, as well as in Nevada's Prida Formation and the Union Wash Formation of eastern California. The Popo Agie Formation of Wyoming and Utah contains the bones of dicynodonts and rauisuchians, as well as the footprints of dinosaurs and other reptiles. Fossils of cycads, conifers, and ferns associated with coal seams are found in the Pekin Formation of North Carolina and Virginia and in the Santa Clara Formation of Sonora, Mexico. Recently, dinosaur, thecodont, turtle, pterosaur, and early mammal bones have been reported from the Fleming Fjord Formation of the east coast of Greenland.

Where Can Triassic Fossils Be Seen in the Field?

While Triassic sediments are relatively uncommon in North America, there are several spectacular preserves where visitors can see *in situ* fossils from this time. All are in the American Southwest. The most famous of these is undoubtedly Petrified Forest National Park east of Holbrook, Arizona, where exposures of the Chinle Formation are known for the colossal fossil trees that they contain. Logs tens of feet long are not uncommon—in fact, most of the logs you'll see along the tourist trails are large. Why? If you'll let me up on my soapbox for just a moment, I'll explain: tourists have pilfered all the smaller pieces, even though theoretically the park is a protected area. It's sad, because the small minority of people who ignore the rules are threatening the park itself with extinction. That actually happened to Fossil Cycad National Monument in South Dakota early in the twentieth century. Within a decade of being designated a park, all the fossil cycads had been stolen, and the monument was eventually decommissioned. Petrified Forest perhaps has some measure of immunity because of the sheer size of the biggest logs, but already what you see on a casual walk is not what the fossil forest is really like. To see that you have to hike into the backcountry, where few tourists go, and there you will find places where the rainbow-colored wood is so plentiful that you cannot help stepping on it. You may even see a phytosaur tooth lying in the dirt or a metoposaur skull weathering out of the rock!

Petrified wood *Araucarioxylon arizonicum*, Chinle Formation, Petrified Forest, Arizona. Denver Museum of Nature & Science, Denver, Colorado.

Two other Triassic preserves are much better-protected merely by virtue of being much smaller and thus more manageable. Also in the Chinle, the Ghost Ranch Quarry near Abiquiu, New Mexico has produced dozens of *Coelophysis* specimens, many of these dinosaurs articulated and complete. The Ruth Hall Museum of Paleontology, on the ranch property, houses exhibits of fossils from the quarry and elsewhere, and the quarry pit itself can be reached by a short hike from the parking lot. Bring a pair of tall boots, though, because the last time I was there, there was a sign on the fence warning you to hike at your own risk because of the possible presence of rodents carrying the plague. The walk is short and on an easy trail without significant elevation gain, so other than footwear, you need no special preparations.

Phytosaur scutes (above) and lungfish palatal tooth plate (below), Dying Grounds, Chinle Formation, Petrified Forest National Park. Fossils like these can only be found in the backcountry where tourists don't go.

Berlin-Ichthyosaur State Park outside of Austin, Nevada protects exposures of the Luning Formation that yield many individuals of the ichthyosaur *Shonisaurus*. Although many bones are scattered, there is a building, the Fossil House, built over a quarry with a large number of bones exposed but still in the ground. It might be wise to inquire ahead about the Fossil House hours, as when I was there some years ago, the place was locked, with nary a ranger to be seen, and I had to peer in the windows. This is a rather out-of-the-way state park and doesn't get the tourist traffic that some other spots do.

I am unaware of any preserves in North America that were created specifically to showcase Triassic marine invertebrates. Your best bet to see ammonoids, brachiopods, belemnites, and corals from this timeframe is to return to our old friend Death Valley National Park and look in exposures of the Butte Valley Formation.

The Triassic had seen the emergence of both the dinosaurs and the mammals, as well as the conquest by tetrapods of the seas and the skies. However, the dinosaurs would dominate the landscape for some 135 million years to come. Our ancestors would hide in the shadows while the "ruling reptiles" ascended to power in the subsequent period, the Jurassic.

Chapter Ten
The Jurassic

Eastern Utah, 140 million years ago: three hungry *Allosaurus* try to bring down a lone *Diplodocus* — an old male. Overhead, a *Rhamphorhynchus* eagerly awaits the chance to snatch scraps. The summer skies hold welcome rain for the cycads and *Ginkgo* trees that dot this seasonally arid Morrison floodplain.

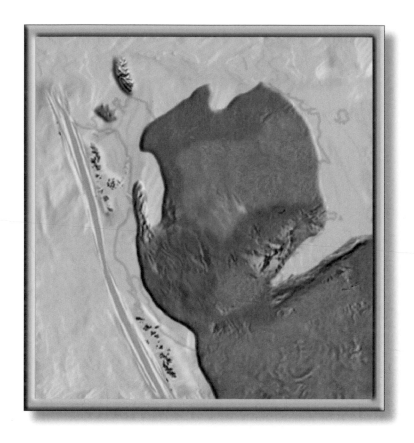

Jurassic North America

he Jurassic was the "high noon" of the Age of Dinosaurs—the middle period of the Mesozoic Era, and the time during which the dinosaurs came to rule the Earth. Almost everyone can imagine a Jurassic landscape, replete with herds of long-necked sauropods browsing on high conifers and low cycads, and both solitary and pack-hunting theropods lurking in the shadows, waiting for a young or old individual to stray from the safety of the herd.

Definition and Nomenclature

The name Jurassic comes from the Jura Mountains, an outlier of the Alps along the French-Swiss border, where rocks of this age were first studied. Alexander von Humboldt chose the moniker in 1795, but it was not officially proposed as a Period name until Leopold von Buch did so in 1839.

No global stratotype sections have yet been designated for either the start nor the end of the Jurassic. The International Commission on Stratigraphy uses the working definition of "near the lowest occurrence of the smooth *Psiloceras planorbis* ammonite group" for the beginning of the Jurassic, which has a radiometric (U-Pb) date of 200 million years ago. Similarly, they use "maybe near the lowest occurrence of the ammonite *Berriasella jacobi*" as the working definition for the Jurassic-Cretaceous boundary (end of the Jurassic), dated at 145 mya. Thus the Jurassic may, for the time being, be considered to have lasted about 55 million years.

Paleogeography and Paleoclimate

We left the Triassic with all the continents agglomerated together in a single landmass, Pangea. This stretched nearly from pole to pole and was cleft in its eastern midregion by the Tethys Sea. North America was firmly sutured to South America through Mexico, and the nascent rumblings of breakup could be heard as an inceptive rift was forming between our East Coast and the eastern continents of Europe and Africa.

The Jurassic was the time during which Pangea seriously began to fracture. Rift valleys, similar to those seen in East Africa today, had formed along North America's Eastern Seaboard during the Late Triassic, creating downdropped valleys (grabens) into which streams deposited their sediments. This activity continued into the Jurassic, with the valleys becoming more pronounced and harboring lakes and, later, the embryonic Atlantic Ocean. As the crust subsided, more than five miles of sediment gradually accumulated in these rift valleys, which are exposed today from South Carolina to Nova Scotia. Buried extensions of the rift lie as far south as Florida and as far north as Newfoundland. Discontinuous exposures of these sediments form the Newark Supergroup, which spans the Triassic/Jurassic boundary.

Sauropod dinosaur Camarasaurus, Morrison Formation, Utah. Museum of Western Colorado, Fruita.

During this time, Pangea continued to drift slowly northward, so that at long last, North America lay almost entirely within the northern hemisphere. Our continent still occupied a position somewhat south of where it is today, such that the northern high-pressure arid belt spanned most of the United States. This led to extensive expanses of deserts, especially early in the Jurassic, including vast areas of sand dunes that have been petrified as the Entrada and Navajo sandstones of the western states. It has been estimated that these dune fields originally exceeded those of the Sahara in size, and may have rivalled those of the Empty Quarter of modern Saudi Arabia. In the east, many of the rift lakes were salty because of the dry, continental climate, although as the Atlantic opened, access to maritime moisture caused rainfall to increase. The monsoonal climate that prevailed across Pangea during the Triassic abated as the breakup of the supercontinent progressed, and rainfall spawned over the ocean gained reliable access to the interior regions of the land.

The Appalachians were still fairly young, and provided significant topographic relief in the east. The Ancestral Rockies had been worn down to low hills that kept the south-central Western Interior above sea level, but not by much. During the early portion of the Jurassic, the Sundance Sea—which was connected on the west and north to both the Pacific and Arctic Oceans—inundated a large area covering much of Utah, Nevada, Idaho, Wyoming, Montana, and Alberta. This sea gradually retreated and left those inland areas dry, but close to sea level. Sediments laid down in the Sundance Sea comprise most of the few Jurassic marine exposures in North America: otherwise, the rock record is almost entirely terrestrial. As the Atlantic slowly opened, the North American plate moved westward and overrode the Pacific plate. This created an offshore trench from California to Alaska, accompanied by strings of volcanic mountains that lay approximately as far inland as the California-Nevada border. Intermittent streams drained these highlands to the east, meandering across a vast low-relief plain that covered much of the interior of the continent during latest Jurassic time.

The greenhouse climate of the Triassic continued into the Jurassic. There was no ice at the poles; in fact, high-latitude climates were only cool-temperate, and the subtropics extended as far as 60°N. In Greenland, temperate forests grew and coal deposits formed. There may have been some snow on the tops of high mountains, but other than that, the world was one of perpetual summer. North America's weather was considerably warmer than it is now, both because of global warmth and because of its more southerly latitude. The equator ran through Central America and the Caribbean.

Life in the Jurassic

Life in the oceans was looking more and more modern. The ubiquitous brachiopods of the Paleozoic were largely replaced by clams. Snails were increasingly common. Crinoids still hung on, but with nowhere near their former glory; urchins and starfish were the more obvious echinoderms now. There was a plethora of cephalopods—the oceans swarmed with ammonites and belemnites, as well as squid and octopi. Reefs were built mainly of sponges and hexacorals. Teleosts appeared, the group of ray-finned fish that contains the majority of modern forms. Sharks, of course, were numerous, but they were not top predators: those roles were filled, as they had been during the Triassic, by the plesiosaurs and ichthyosaurs, along with marine crocodiles. So, whereas the marine invertebrate suite was taking on a modern look, the maritime menagerie was still very different from that inhabiting the modern oceans.

There wasn't much of a change in terrestrial vegetation from the Triassic to the Jurassic. The flora was dominated by gymnosperms, with the most common forest trees being conifers of the araucaria and podocarp types. A botanically-savvy visitor would recognize these as relatives of the Norfolk Island Pine and Buddhist Pine of today's Australasian region, but the look would have been quite different from that of a

Eubrontes (theropod dinosaur) track, natural cast, western Colorado. Museum of Western Colorado, Fruita.

modern boreal forest. Ginkgos (maidenhair trees) flourished, especially in middle to high northern latitudes; "shrublands" were mostly dominated by cycads and cycadeoids. There is no fossil evidence that angiosperms (flowering plants) had evolved yet, although molecular genetic studies suggest that their roots may go back into the Jurassic. Herbaceous ground cover was mostly ferns, clubmosses, and horsetails, the former of which grew to the size of small trees in habitats with abundant moisture. There was no grass, as it is classified within the flowering plants.

The cast of animals bore little resemblance to today's fauna. Mammals were still small, ratlike creatures relegated to skulking in the shadows. A variety of archaic groups with tongue-twisting names like haramiyids, amphilestids, docodonts, morganucodontids, triconodonts, and multituberculates lived in the wake of the dinosaurs, eking out a living eating mostly insects. The smallest were only a couple of inches long, and the largest were about the size of a modern opossum, with most the size of a mouse or a rat. Multituberculates ('multis') were the only group whose teeth clearly indicate an omnivorous diet—the name comes from the pattern of cusps or tubercles on their molars, which were set in two or three parallel rows and resemble the tread on a mountain bike tire. The multis were the group that was most numerous and diverse, and in contrast to the others which were primarily confined to the Jurassic, the multis persisted all the way into the Oligocene, some 100+ million years later; they were, in fact, the only Mesozoic mammal order to survive into the Tertiary. All of these mammals had a heterodont dentition (one with several different types of teeth): anterior incisors, stabbing canines, and crushing or chewing premolars and molars. As we have seen, fossil evidence of pits for the innervation of whiskers indicates that they possessed fur, which leads to the conclusion that they were both warm-blooded and nursed their young. Their brain sizes weren't yet much improved over their therapsid ancestors, however, and the structure of the jaw hinge was still archaic—whereas most Jurassic mammals possessed a joint between the dentary and squamosal bones (like modern mammals do), they also retained the primitive articular-quadrate joint they had inherited from their reptilian forebears.

But the landscape increasingly belonged to that group of diapsid reptiles that had made their debut during the Triassic: the dinosaurs. Various "terrible lizards" filled almost every terrestrial econiche. We have already discussed how dinosaurs may be divided into saurischian (lizard-hipped) and ornithischian (bird-hipped) types, based upon the orientation of the pubic bones. (One theory as to why some dinosaurs had a posteriorly-oriented pubis is that it allowed for a larger

Snails
Viviparus reeside,
Morrison Formation, Colorado. Denver Museum of Nature & Science, Denver, Colorado.

gut, which would have been of particular advantage for herbivores.) It was mainly the saurischians who dominated the Early Jurassic landscapes. This group split into two lineages, the herbivorous prosauropods and sauropods, and the carnivorous theropods. The prosauropods, which had gotten a good start during the Triassic, soon spawned that quintessential dinosaur group—the sauropods: those long-necked, long-tailed, small-headed herbivores familiar to anyone who has ever seen a Sinclair Oil sign. At first, theropods were small, but they soon grew to gigantic proportions. The most familiar Jurassic theropod is *Allosaurus*, but there were others built upon the same theme, such as *Ceratosaurus* and *Dilophosaurus*. All were bipedal (two-legged) carnivores.

Ornithischian dinosaurs got off to a slower start. Small, generalized fabrosaurs and heterodontosaurs roamed the landscape as they had done during the latest Triassic, and gave rise to the first iguanodontids; but the most familiar Jurassic ornithischian group is the stegosaurs. These large, plated herbivores need no introduction. Their double, staggered row of triangular plates, embedded in the skin along the vertebral column, is known to every schoolchild. They may, in fact, be the most recognizable of dinosaurs. For many years it was thought that the plates were used as protection against the large, predatory theropods, but this theory is no longer in vogue. The recognition of canals for an ample blood supply suggests that the plates were used instead as thermoregulatory devices, similar to the 'sails' of Permian pelycosaurs like *Dimetrodon*. By facing towards or away from the sun, as need be, these plates could act like a set of solar panels or an automobile's radiator, helping to keep the dinosaur's body temperature within an optimum range. A set of spikes on the tail could be wielded for protection against a hungry *Allosaurus*.

This brings us to the question of dinosaurian metabolism. All the hoopla surrounding the discovery of a bevy of feathered dinosaurs in China during the last decade has led most scientists to accept that at least some dinosaurs possessed an endothermic (warm-blooded) metabolism. The boldest will therefore assert that all dinosaurs were warm-blooded, but this is jumping the gun a little. All of the feathered dinosaurs are theropods, the group from which birds arose. Clearly, a dinosaur is not a dinosaur is not a dinosaur—they were a very diverse group in size, habitat, and lifestyle. Why should it follow that if some theropods evolved feathers for insulation, a group as far removed from them as sauropods or stegosaurs would possess a high metabolism as well? In a climate of perpetual summer, a large animal can stay warm merely by virtue of its low ratio of surface

Adult & juvenile stegosaurs *Stegosaurus stenops*. Adult - Albany Co., Wyoming; juvenile - Uinta Co., Utah. Denver Museum of Nature & Science, Denver, Colorado.

area to volume—an endothermic metabolic rate is neither necessary nor advantageous, and energetically costly in terms of required food intake. Elephants, in fact, are troubled to rid themselves of excess metabolic heat in their tropical homeland (witness the large ears, which function as radiators), and studies have shown that an animal the size of a large dinosaur could easily maintain an equable body temperature in Mesozoic climates with a metabolic rate no higher than a crocodile's. Such a condition is called inertial homeothermy, and is likely to have been the condition amongst most large dinosaurs. The occasional sauropod skin fossil, showing a bald hide studded with small armor scutes, supports such a conclusion. Additional studies of the ratio of brain to body size in various modern animals show that a large relative brain size tends to correlate with a high metabolism; amongst dinosaurs, theropods have brain sizes the closest to the mammalian/avian range, whereas sauropods and stegosaurs have encephalization ratios squarely in the reptilian portion of the graph—another piece of evidence suggesting that metabolic rates differed amongst different types of dinosaurs.

Small theropods, however, would have benefited from maintaining a high activity rate in pursuing their prey, and their size would not have given them a surface-to-volume ratio that was efficient at retaining heat. It is among these dinosaurs that we find evidence of feathers. It is tempting to assume that feathers mean insulation, which consequently means they were trying to keep warm; but it is also possible that feathers first evolved for display purposes, and were only later co-opted for their insulative value. Nevertheless, it is no accident that this is the lineage from which the first birds are derived, and no one would argue that birds aren't warmblooded. This is highly suggestive of endothermy predating the dino-bird split.

Cycadeoid trunks: above - whole; below - sectioned. Provenance unspecified. University of Michigan Exhibit Museum of Natural History, Ann Arbor.

That birds, which first appear during the Jurassic, evolved from small theropod dinosaurs is nigh unto indisputable. The skeleton of *Archaeopteryx*—the first known bird—so resembles that of the dinosaur *Compsognathus* that for a hundred years a specimen of the former was misidentified as one of the latter. *Archaeopteryx* retains a mouthful of teeth, clawed fingers, and a long, bony tail—all dinosaurian features—as well as a semilunate carpal bone (in the wrist), which links the claw-slashing behavior of the group of theropods known as maniraptorans ("raptors" in common parlance) with the wing's flight stroke in birds. The asymmetrical shape of *Archaeopteryx'* feathers, indistinguishable from the construction of the flight feathers of modern birds, indicates that *Archaeopteryx* was fully capable of powered flight. By the Late Jurassic, therefore, pterosaurs no longer had the skies to themselves. As with pterosaurs, whether avian flight originated from the ground up or the trees down remains a matter of debate.

The Morrison Formation, the main Late Jurassic sedimentary unit in western North America, has been extensively mined for paleoecological information. It covers a vast area of the Western Interior stretching from New Mexico to Alberta and Saskatchewan. Sedimentary evidence indicates that the Morrison was a dry floodplain, at least seasonally, especially in its southern ranges. It has been compared to the climate of Kenya's Amboseli, but the 'savannah' vegetation would have been entirely different in the absence of grasses. Cycads, ginkgos, and conifers formed gallery forests along streams, some of which may have been perennial and others seasonal. Crayfish burrows 5 m (16 ft) deep in the Colorado Plateau region indicate that the permanent water table must have been no deeper than that; thus tree roots could have reached groundwater even when a stream was not flowing. During late Morrison time, a large, shallow, saline lake—Lake T'oo'dichi'—with an areal extent approximately the same as that of Lake Michigan covered portions of New Mexico, Arizona, Utah, and Colorado. The fact that this lake was salty indicates that the climate was at least seasonally dry. What vegetation may have grown as ground cover in drier areas is a speculative matter—ferns were the most common herbaceous plants during the Jurassic, but most ferns require a moister environment than is present in a savannah climate. Many cycads have tuberous stems and could survive until seasonal rains arrived, and are thus good candidates for interfluvial vegetation.

Further north, in Montana and the Canadian prairie provinces, the Morrison environment seems to have been wetter, with more reliable moisture. This makes sense with the location in the midlatitudes, north of the aforementioned 'belt of aridity.' Here there were probably more permanent streams and lakes than further south in Colorado and Utah. Fossils of cycadeoids, conifers, ginkgos, seed ferns, and true ferns have come from the Morrison in Montana and southern Canada. Dinosaur bones, paradoxically, seem to be equally plentiful in both the drier and wetter regions, which may indicate that the herds migrated as they followed the lush vegetation that sprouted after seasonal rains. Now that the old-school vision of swamp-dwelling sauropods has fallen

Sauropod dinosaur *Camarasaurus* skull, Morrison Formation, Utah. American Museum of Natural History, New York.

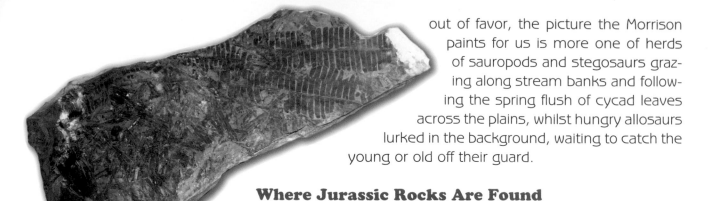

out of favor, the picture the Morrison paints for us is more one of herds of sauropods and stegosaurs grazing along stream banks and following the spring flush of cycad leaves across the plains, whilst hungry allosaurs lurked in the background, waiting to catch the young or old off their guard.

Where Jurassic Rocks Are Found

Cycadeoid leaf *Zamites arcticus*, Morrison Formation, Colorado. Denver Museum of Nature & Science, Denver, Colorado.

Thin bands of Jurassic sediments are exposed all along the Eastern Seaboard from South Carolina to Nova Scotia. Most of these rocks are those laid down in the rift valleys discussed earlier, as Europe and Africa began to unzip from North America, giving birth to the Atlantic Ocean. Jurassic rocks are essentially nonexistent in the midcontinent, although there are small areas of Jurassic bedrock in Iowa and central Michigan buried beneath a mantle of glacial till.

The most extensive Jurassic exposures are in the Western Interior, where terrestrial sediments are exposed over vast areas of New Mexico, Colorado, Utah, Wyoming, Montana, Alberta, and Saskatchewan, spilling over a bit into neighboring states. Northern California and mountainous areas of British Columbia and Alaska also harbor considerable Jurassic outcrops. Smaller exposures are found in an east-west band through southern Mexico and on into Guatemala and Nicaragua. Cuba also hosts some Jurassic rocks on the western end of the island.

Where Jurassic Fossils Are Found

The Jurassic fossil record of eastern North America is largely one of footprints. The rift valleys that formed as Pangea split apart contain a plethora of sediments that preserve dinosaur tracks, running all up and down the East Coast from the Carolinas to Massachusetts. Most of these are theropod tracks given names like *Eubrontes* and *Grallator*, but there are also other tracks such as those of prosauropod dinosaurs and crocodiles. The Connecticut River valley is one of the richest sources of dinosaur footprints in the world, with

Unidentified dinosaur eggs, Morrison Formation, Cañon City, Colorado. Denver Museum of Nature & Science, Denver, Colorado.

tens of thousands of tracks identified since the first were uncovered by a farmer in 1802. (Surprisingly, bones are rare in the eastern formations, although a few nice specimens have been found.) Most of these tracks are in the Newark Supergroup in sediments of the East Berlin, Shuttle Meadow, and Portland formations—although Nova Scotia's McCoy Brook Formation preserves the best overall suite of footprints from anywhere on the continent. When tracks are found but bones aren't, it's a lot of detective work trying to figure out who made the footprints. Of course, knowing which dinosaurs lived at the same time and had a foot shape capable of making a particular track puts one well on the trail of the perpetrator. It takes a different type of a person to marvel as much at a dinosaur footprint as at a humongous skeleton in a museum, but in their own way, they are just as fascinating—and have sometimes been labelled "snapshots of behavior."

In the West, the dinosaurian story is predominantly told by the Morrison Formation—although the rocks also reveal the presence of turtles, crocodylians, lizards, sphenodontians, and mammals. The underlying eolian Kayenta Formation, Navajo Sandstone, and Entrada Sandstone preserve a few footprints but are generally depauperate of bones; but these pale in comparison with the osteological riches of the Morrison. The Morrison contains the richest concentration of dinosaur bones in any single formation anywhere. Thousands of bones from hundreds of individual dinosaurs (including many articulated skeletons) have been found in this formation. The beds are a distinctive multicolored series of reddish, greenish, and greyish sandstones, mudstones, and shales which produce an array of vegetarians like *Apatosaurus*, *Camarasaurus*, *Diplodocus*, and *Stegosaurus* as well as meat-eaters like *Allosaurus*. Much of the digging that accompanied the Cope-Marsh 'bone wars' of the late nineteenth century occurred in the Morrison Formation.

Sauropod dinosaur *Diplodocus longus*, Morrison Formation, Uinta Co., Utah. Denver Museum of Nature & Science, Denver, Colorado.

South of the border, exploration has not been as extensive, but mid-Jurassic footprints have been reported from the state of Michoacan in southwestern Mexico, and scattered evidence of other dinosaurs comes from elsewhere in the country. As time goes on, we can expect fieldwork to reveal more exciting dinosaur discoveries from our southern neighbors.

Where Can Jurassic Fossils Be Seen in the Field?

You're in luck if you want to see big Jurassic dinosaur fossils—there are 'tons' of places to go and view them in place. If you live in, or travel to, the east, it's going to be

Petrified wood cf. *Araucarioxylon*, Zuni, New Mexico. Private collection.

footprints that you see. The west has footprints too, as well as big bones.

Dinosaur State Park, in Rocky Hill, Connecticut, showcases scores of dinosaur tracks in the Shuttle Meadow, East Berlin, and Portland Formations. One likely candidate for the maker of the tracks displayed here is *Dilophosaurus*, the familiar crested theropod from the first *Jurassic Park* movie and novel (although there is no evidence whatsoever that this dinosaur really could spit poison). Another famous set of trackways is at Mt. Tom near Holyoke, Massachusetts, where 26 parallel *Grallator* trails indicate that a herd of dinosaurs moved through the valley together.

Sparse as Jurassic exposures are in the midwest, there are dinosaur tracks to be seen in the Morrison Formation at Black Mesa State Park near Kenton, Oklahoma. This is about as far east as outcrops of the Morrison go. Get a little further west, however, and it's Morrison, Morrison, Morrison.

Premier among the Morrison parks is Dinosaur National Monument, which straddles the Colorado-Utah border near Vernal, Utah. This has been called "the greatest dinosaur bonebed ever discovered in North America," and houses a splendid visitor center built onto the side of a hill where hundreds of dinosaur bones, including articulated skeletons of several *Camarasaurus* individuals, have been excavated and left exposed in relief in the rock. Some years ago you could visit and actually watch paleontologists in slings chipping away at the bones, but now the excavation work has been finished. The bonebed represents stream deposits that concentrated dinosaur skeletons, mostly sauropods, at this particular spot in the river. Also near Vernal is Red Fleet State Park, where dinosaur tracks are exposed along the shore of the reservoir, this time in the Carmel Formation.

Theropod dinosaur *Allosaurus fragilis*, Morrison Formation, Utah. Museum of Western Colorado, Fruita.

Further south in Utah, a fair drive east of Price on gravel roads, is another spectacular Morrison quarry: Cleveland-Lloyd. It's not as fancy as Dinosaur National Monument, but it gives visitors a chance to see other species of dinosaurs. The Cleveland-Lloyd Quarry has a more rustic shelter and museum but also exhibits an *in situ* display of bones, this time with a predominance of *Allosaurus* individuals. The scenario here is thought to have been one of predators drawn to prey around a drying water hole.

A small park with easy access for visitors with limited time or physical ability is Rabbit Valley, in westernmost Colorado. This exit off I-70 just a few miles from the Utah border has a short loop walking trail through Morrison Formation rocks, where you can see several large boulders with dinosaur bones entombed.

The Mill Canyon Dinosaur Trail, near Moab, Utah, takes you past still-in-place dinosaur bones. If you find yourself in southern Utah, you should also check out Glen Canyon National Recreation Area and Grand Staircase-Escalante National Monument. The former harbors large petrified logs along with dinosaur tracks in exposures of the Navajo Sandstone, whereas the latter's Entrada Sandstone is home to a plethora of theropod and sauropod tracks—though you'll have to hike cross-country to see them. Horseshoe Canyon in Canyonlands National Park has similar exposures of the Navajo with more fossils like those at Glen Canyon.

Another protected, but rather out-of-the-way, area where you can hike to some footprints is Bighorn Canyon National Recreation Area in north central Wyoming. Outcrops of the Sundance Formation preserve the tracks of theropod dinosaurs, as well as some more unusual prints attributed to pterosaurs. Nearby, close to Shell, Wyoming, is the Red Gulch Dinosaur Tracksite on Bureau of Land Management (BLM) land. This site, also in the Sundance Formation, has a wheelchair-accessible walkway from which to view the tracks.

Not actually a park, but nonetheless a stop no traveller should miss, is the Bone Cabin east of Medicine Bow, Wyoming. Back in the late nineteenth century, ranch hands built a small cabin entirely out of pieces of dinosaur

Theropod dinosaur tracks cf. *Grallator*, Newark Supergroup, Connecticut River Valley. Private collection.

bone that littered the local landscape, mortared together like cobblestones. This remarkable little cabin still stands, and houses historical artifacts. Go into the tourist shop next door and pay a small fee, and they'll unlock it for you and let you look around.

The Jurassic gave way to the Cretaceous without any catastrophic changes. The climate remained warm, and new types of dinosaurs evolved from the old. The 'ruling reptiles' were at their zenith and would remain so for another 80 million years. Sea levels gradually rose as plate tectonic activity increased, such that in many areas there was a change in rock types to the chalky marine sediments that characterize the Cretaceous. Next, it is to these that we will turn.

Ornithischian dinosaur *Stegosaurus stenops*, Morrison Formation, Colorado. Natural History Museum of Los Angeles County, Los Angeles, California.

Chapter Eleven
The Cretaceous

For much of the Cretaceous, the Western Interior Seaway cleaved our continent in two, stretching from the Gulf of Mexico to the Arctic Ocean. This Colorado landscape portrays the western shore of that sea, where *Tyrannosaurus rex* casts his eye on a young hadrosaur. Three pterosaurs soar overhead, while a pliosaur basks on the sand, ready to plunge to safety if need be.

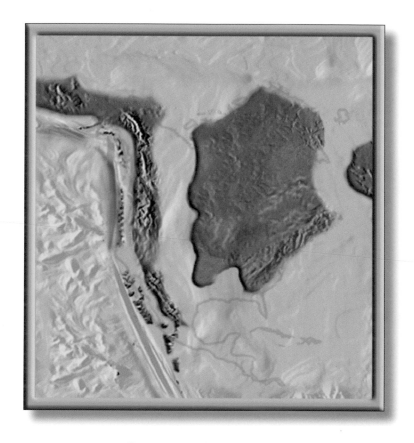

Cretaceous North America

We come now to the third and final period of the Mesozoic Era. The transition from the Jurassic to the subsequent Cretaceous was a quiet one, marked in North America, as elsewhere, by a change in rock types from largely terrestrial sediments to an abundance of marine chalks. The period ended far from quietly, however—as we shall soon see.

Definition and Nomenclature

In 1822, Belgian geologist d'Omalius d'Halloy proposed the name Cretaceous for the chalky sandstones and shales exposed in and around France's Paris Basin, and similar deposits in neighboring Belgium and Holland, most famously visible at England's White Cliffs of Dover. The moniker comes from 'creta,' the Latin word for chalk; in German, the word is 'kreide,' from whence comes the abbreviation for the Cretaceous, K—thus the term 'K/T boundary' for the dividing line between the Cretaceous and the following period, the Tertiary. We will discuss the K/T boundary in quite some detail later. First, let us define the Cretaceous. Its beginning, or the Jurassic-Cretaceous boundary, is yet to be tied to a global stratotype section, but its working definition is "maybe near the lowest occurrence of the ammonite *Berriasella jacobi*." The most current dating places this at 145 million years ago. The K/T boundary, or end of the Cretaceous, was defined in 1991 at the iridium geochemical anomaly at the base of the boundary clay, El Kef, Tunisia, dated to 65 mya. The Cretaceous may thus be considered to span some 80 million years.

Paleogeography and Paleoclimate

For some time we have watched the peregrinations of our continent as it left equatorial latitudes in the late Paleozoic and drifted steadily northward throughout the Mesozoic, rotating slowly counterclockwise as it did so. By Cretaceous times, North America was at long last entirely situated within the hemisphere it occupies now. The breakup of Pangea continued as the Atlantic unzipped from south to north and slowly separated us from Europe and Africa. At the same time, South America failed to keep pace with North America on its journey northward, and the Caribbean opened. The southern edge of our continent lay approximately along the Mexico-Guatemala border, and for the first time volcanoes appeared in Cuba and the Antilles. Worldwide, the modern continents were taking shape, and by the Late Cretaceous, the map of the globe we are familiar with today was recognizable, differing only in detail. The equator skirted the northern shore of South America.

Pterosaur *Pteranodon ingens*, Logan Co., Kansas.
American Museum of Natural History, New York.

The warm, equable climate the globe had enjoyed since the dawn of the Mesozoic continued, with climatic zones displaced some 15° poleward of where they lie now, and still only cool-temperate climates (no ice) in the high northern and southern latitudes. Fast rates of plate tectonic movement (about three times what they are currently) resulted in sea levels some 200 m (650 ft) above today's by the mid-Cretaceous, flooding about a third of North America's land surface—the most extensive inundation since Ordovician times. These epicontinental seas were most notable in the midcontinent, where the Western Interior Seaway stretched across the low-lying plains to connect the Arctic Ocean with the Gulf of Mexico. This effectively split our continent into 'West North America' and 'East North America' for much of the Cretaceous. Large portions of what are now the Gulf Coastal and Atlantic Coastal plains were also inundated, while the Appalachians remained above water. Only southern Mexico retained the dry climate of the 30°N arid zone; maritime influence was such that most areas of the United States enjoyed a moister climate overall than they had during the Jurassic. Southern Canada was warm-temperate and the Arctic isles cool-temperate, including Alaska, which intruded above the Arctic circle for the first time.

Ammonite *Sphenodiscus lenticularis*,
Fox Hills Formation, South Dakota.
Private collection.

It was in these warm, shallow seas that the great Cretaceous chalk formations were laid down. The chalk is formed from countless coccoliths—skeletal plates of microscopic marine algae. These algae live only in warm, clear waters such as existed along the eastern side of the Western Interior Seaway. West of the Seaway, the Rockies were rising, and sediments eroded from them resulted in turbid waters, excluding the coccolith algae and precluding chalk formation. Deep ocean waters may have been some 8° to 11°C (15° to 20°F) warmer than they are today, and the deposition of black shales indicates that bottom waters were hypoxic to anoxic (low in, or devoid of, oxygen) in many places. Instead of ocean circulation being primarily temperature-driven (like it is now), weak salinity gradients and widespread stagnation were the norm.

Life in the Cretaceous

We have noted that by Jurassic times, ocean life was becoming more and more recognizably modern. This trend continued throughout the Cretaceous. Sharks and teleost fish were the most numerous marine vertebrates, as they are now; however, the niche of top predator was still filled by the various types of marine reptiles. Both long-necked and short-necked plesiosaurs remained common. The great marine crocodylians were on the wane; so were the ichthyosaurs, which went extinct in the Late Cretaceous, millions of years prior to the K/T boundary. The suggestion has been made that ichthyosaurs could not keep up with the evolution of increasingly fast-swimming prey fish, although this is a difficult hypothesis to test. With the demise of the ichthyosaurs came the rise of the mosasaurs, a group of marine lizards closely related to the terrestrial monitor lizards. Mosasaurs were the dominant marine predators during the latest Cretaceous, sharing the role with humongous carnivorous fish such as *Xiphactinus* (see photo of "fish-within-a-fish" on p. 158). That mosasaurs liked to dine on ammonites is shown by the fossils of many shells of the latter bearing the puncture marks of rows of mosasaur teeth.

Unidentified angiosperm leaves, Meeteetse Formation, Wyoming. Denver Museum of Nature & Science, Denver, Colorado.

Invertebrate life continued much as it had, with the appearance of few notable new groups other than the diatoms (microplankton with siliceous shells). Ammonites and belemnites flourished, as did clams, snails, sea urchins, starfish, and hexacorals. Although scattered evidence of predatory snails dates back to the Cambrian, the Cretaceous fossil record shows a marked increase in this type of attack (drill holes in clam shells). A general increase in predation pressure led to the decline of many benthic (bottom-dwelling) animals and a proliferation of infaunal (burrowing) lifestyles among

marine invertebrates. One new notable group that appeared during the Cretaceous was the rudist bivalves. These 'jumbo clams' were major reef formers throughout the period, but failed to survive the K/T boundary.

The skies played host to both pterosaurs and birds. The toothed, crestless, long-tailed rhamphorhynchoid pterosaurs of the Jurassic were replaced by toothless, tailless, crested pterosaurs called pterodactyloids. But birds evidently gave them some stiff competition, because by the latest Cretaceous, few pterosaurs remained; notable exceptions were the familiar, large-crested *Pteranodon* and the biplane-sized *Quetzalcoatlus*. Most Cretaceous birds were clearly distinguishable from modern birds, particularly by their retention of teeth. They radiated into a variety of econiches, including forest canopies and seashore/diving habitats. Whereas pterosaurs failed to survive the end-Cretaceous event, the fossil record makes it clear that they were declining in both diversity and abundance for many millions of years beforehand.

Sea urchin *Hemiaster whitei*, Goodland Formation, Texas. Burpee Museum of Natural History, Rockford, Illinois.

The fragmentation of Pangea led to the cessation of cosmopolitanism in terrestrial faunas. With the separation of the continents, it became increasingly difficult for land animals to migrate from one end of the Earth to the other. Consequently, evolution took different pathways on different continents, and endemic faunas emerged. One particularly striking example of this may be seen in the success of sauropod dinosaurs in North America vs. in South America. The Cretaceous saw a sharp decline in sauropod diversity and abundance in North America, whereas these long-necked, long-tailed gargantuans remained important herbivores in South America throughout the period.

Why might that be? Is it only the vagaries of evolution, or is there more to the story? Let us examine a particularly telling connection—that of these dinosaurs with their likely food supply. One of the last true revolutions in the evolution of terrestrial life occurred during the Cretaceous—the appearance of angiosperms, or flowering plants. This is the last major group of plants to appear in the fossil record. The first angiosperms are known from the earliest Cretaceous, but it took until the middle of the period before they diversified substantially and began to nudge the gymnosperms out of some of their preferred habitats. Early flowering

Sectioned ammonite *Sphenodiscus* sp., Guadalupe, Chihuahua, Mexico. Note the internal chambers infilled with calcite crystals. Author's collection.

plants were 'weedy' species, small herbs thought to have been successful mainly in successional and disturbed environments. For whatever reason, angiosperms were quicker to gain a foothold in low and midlatitude North America than they were in comparable regions of South America; a tree habit, and their invasion of the forest canopy, also appeared first in the northern hemisphere. The large sauropods, adapted to high browsing of coniferous trees, may have seen their preferred food supply decline in North America, while it persisted south of the equator—and their competitive success in the two regions followed suit.

During the first half of the Cretaceous, these weedy angiosperms were generally a minor component of the flora; conifers, cycads, cycadeoids, and ginkgos remained the dominant forest trees, with an understory of horsetails and ferns. Gradually, though, the flowering plants gained a foothold in a wider variety of environments, first in the low latitudes and progressing steadily northward. This pattern of colonization has been attributed to the slow development of cold tolerance in flowering plants. Concomitant with their rise came the appearance and diversification of several new types of insects, such as the lepidopterans (butterflies and moths) and the eusocial bees. In today's world, these insects are important pollinators of flowers, and it is tempting to believe that the parallel success of the two groups is a pristine example of coevolution. By the latter half of the Cretaceous, flowering plants increasingly came to dominate a wide variety of habitats at the expense of cone-bearing and sporing plants. Mid-Cretaceous angiosperms included small trees, and by the end of the period, forests of primitive magnolias, oaks, elms, poplars, sassafrases, palms, and other familiar trees were common everywhere except at the highest latitudes—where conifers retained an edge, as they do today. Although the first fossil grasses appear toward the end of the Cretaceous, grasslands as an ecosystem did not become established until well into the Tertiary.

Ceratopsian dinosaur Styracosaurus albertensis, Red Deer River, Alberta. American Museum of Natural History, New York.

We have already mentioned how this floristic change may have influenced the evolutionary trajectory of sauropod dinosaurs. What about its effect on other terrestrial animals? The appearance of all three modern mammalian stocks (monotremes, marsupials, and placentals) by the end of the Cretaceous, in addition to the already-successful multituberculates, may be tied to the appearance of new food resources supplied by angiosperms—particularly nectar and fruits. In the shadow of the great dinosaurs, all of these mammals remained small, but their diversity would later be the key to their success. Monotremes—egg-laying mammals represented today only by one species of platypus and two species of echidna from Australia and New Guinea—have a poor fossil record, but our knowledge of marsupials and placentals is much better. Marsupials, whose young are born at a relatively undeveloped stage and spend weeks or months maturing in the mother's pouch, were a highly successful group in the Late Cretaceous and much of the Tertiary, primarily (though not exclusively) on the southern continents; they comprise the vast majority of the endemic mammals of Australia today (e.g., kangaroos, wombats, koalas). Placental mammals, whose young are born at a more developed stage and are generally considered the most advanced of the three types, have been the commonest mammals on the northern continents since Cretaceous times.

New types of dinosaurs also appeared to exploit the new food resources provided by the flowering plants. As the great sauropods and stegosaurs of the Jurassic declined in North America, their place was taken by more advanced dinosaurs with better chewing abilities. Rather than simply strip leaves from branches and grind them with swallowed stones (gastroliths) in a gizzard, as the sauropods had done, these new varieties of herbivores were adapted for effectively chewing their food in the mouth. Cheek pouches to retain food while ruminating and tooth batteries capable of shredding and grinding plant matter made them efficient vegetation-processing machines, and are likely clues to their success. Cretaceous dinosaur faunas were dominated by herds of hadrosaurs or "duck-billed" dinosaurs. Nesting grounds uncovered in Montana and elsewhere indicate that these dinosaurs were social beasts, and that the parents spent considerable time and effort caring for their young. Bonebeds in Alaska and Alberta have been interpreted as indicating that some hadrosaurs, such as *Edmontosaurus*, may have seasonally migrated between northern, summer feeding grounds and southern, winter habitats. The ceratopsians, or horned dinosaurs, are also quintessential Cretaceous behemoths, although in North America, they are found only in the west (remember that for most of the Late Cretaceous, eastern and western North America were isolated from one another by the Western Interior Seaway, which apparently these dinosaurs were unable to cross).

Another new group of herbivores to appear during the Cretaceous was the bone-headed pachycephalosaurs, and the armored ankylosaurs now became quite common. New theropod dinosaurs like *T. rex* also evolved, although most of these were more or less variations on the same bipedal carnivore themes

Theropod dinosaur *Tyrannosaurus rex*, Big Dry Creek, Montana. American Museum of Natural History, New York.

pioneered during the Jurassic. Nevertheless, they were highly successful predators, and their serrated, steak-knife teeth were deadly. Small pockets at the base of the serrations on some theropod teeth (e.g. *Dryptosaurus*) may have been breeding grounds for toxic bacteria, designed to assist in killing by transmitting virulent infections that would cause rapid shock and death in a prey animal that managed to escape. Such a strategy is employed today by the world's largest living lizard, the Komodo Dragon.

A Cretaceous Vignette

Crab *Avitelmessus grapsoideus*, Ripley Formation, Mississippi. Denver Museum of Nature & Science, Denver, Colorado.

A typical Late Cretaceous scene in midlatitude North America, say in Montana near the shore of the Western Interior Seaway, might be an undulating plain tinted green by a mantle of ferns and herbaceous flowering plants, with the higher ground bearing disconnected stands of fig, magnolia, and palm trees. The air is soft and warm, with a light breeze, but the western horizon holds storm clouds. A lone crested pterosaur soars overhead. Waves gently lap the muddy shore, as a flock of toothed shorebirds is probing the beach for oysters exposed by the receding tide. The dry ground further inland hosts a breeding colony of hadrosaurs, their nests full of broken eggshells and wide-mouthed babies squabbling to be fed. A solitary *Triceratops* browses on cycads near the edge of the forest. Two multituberculates chase each other up the trunk of a fan palm, one defending its nest in the crown thicket against the other. Right now, no predators can be seen, but one never knows when the shrubby sassafrases might part and a pack of *Deinonychus* appear.

The K/T Extinctions

Few readers of this book will be unfamiliar with the fact that the K/T boundary marks one of the greatest mass extinctions of all time. Some 70-85% of all species perished, including all land animals with a body weight in excess of 25 kg (55 lbs). It is commonly stated that "the dinosaurs" went extinct at the end of the Cretaceous; this is not strictly true, given the fact that birds are the descendants of—and, cladistically speaking, truly are—small theropod dinosaurs. Nevertheless, all the gigantic, majestic creatures our imaginations conjure up when thinking of a dinosaur did die out at the end of the period, along with all remaining pterosaurs, mosasaurs, plesiosaurs, ammonites and belemnites, and the rudist

Unnamed cycad, Almond Formation, Rock Springs Uplift, Wyoming. Collections of the Denver Museum of Nature & Science, Denver, Colorado.

bivalves. Placental mammals survived relatively unscathed, as did birds, crocodiles, turtles, frogs, salamanders, snakes, nautiloid cephalopods, clams, and snails. Marine fish, lizards, marsupial mammals, and many microplankton (diatoms, dinoflagellates, foraminifera) suffered heavy losses but managed to pull through. What was going on? How can we explain this pattern of extinction?

The evidence is nigh unto indisputable that 65 million years ago, a gigantic meteorite some 10 km (6 miles) in diameter slammed into the Earth, bringing curtains to the Cretaceous world. The first hint of this came in 1980, when a team of researchers headed by Luis and Walter Alvarez reported the discovery of high concentrations of the rare platinum-group element iridium in a clay layer marking the K/T boundary at Gubbio, Italy. Iridium is rare in the Earth's crust, but is far more common both in meteorites and in the Earth's mantle. They suggested that the abundance of iridium in the boundary clay—some 30 times what would otherwise be expected—indicated the fallout from the impact of an extraterrestrial visitor, which was responsible for the mass extinction of the (nonavian) dinosaurs and other aforementioned creatures. This impact purportedly caused 1) an initial frying of all life in the vicinity; 2) global forest fires; 3) a dust-induced global cooling, then 4) a carbon-dioxide-induced global warming, coupled with 5) widespread acid rain—resulting in total ecosystem collapse at the end of the Cretaceous.

Whereas the public embraced this catastrophic theory straight away, the scientific community was slower to get on the bandwagon. Could not massive volcanism have also supplied the iridium? Doubters pointed to the Deccan Traps, huge basalt flows in India and Pakistan dated to just the same time. (You may recall that other flood basalt episodes have been implicated in the end-Permian and end-Triassic catastrophes.) But evidence for the impact theory slowly mounted throughout the decade of the 1980s.

Fish-Within-A-Fish

The boundary clay layer with an iridium spike was located and analyzed at over a hundred sites worldwide. It was found to contain two additional strong indicators of an extraterrestrial impact: microtektites and shocked quartz. Tektites are small spherules of glass which form when molten rock is explosively introduced into

Fish-within-a-fish: Xiphactinus with remains of its last meal, Niobrara Formation, Kansas. Sternberg Museum of Natural History, Hays, Kansas.

the atmosphere and cools rapidly. Alone, they do not prove an impact, as they may also form during explosive volcanism. But the "shocked" quartz was another matter. These are grains of quartz which reveal multiple sets of parallel lamellae along which slippage within the crystal lattice has occurred. Volcanism can cause slippage along a single plane, but has never been shown to produce "shocking" in multiple directions—and some K/T shocked quartz grains exhibit slippage along as many as eight multidirectional planes. The only known phenomena which can produce this type of slippage are extraterrestrial impacts and man-made nuclear explosions. For obvious reasons, the latter is out of the question, leaving us with a strong case for meteorite impact. Since that time, tsunami ('tidal wave') deposits have been recognized at Brazos River, Texas and elsewhere in the Gulf of Mexico region, pointing to an impact location in the Caribbean somewhere. A spike in fern spores appears just above the boundary clay in the Raton Basin straddling the Colorado-New Mexico border; this is thought to indicate recolonization following a total vegetational dieback. Supporters point to historical accounts of the devastating eruption of the island volcano of Krakatoa in Indonesia in 1883: ferns were the first plants to recolonize the island following the eruption. Even today, ferns are the

The K/T boundary clay can be clearly seen in the Raton Basin of southern Colorado; it is the white layer my mom is pointing to. Near I-25 exit 8.

first colonizers of fresh lava flows from the Kilauea eruptions on the island of Hawaii. Nevertheless, all of this evidence was circumstantial, and doubters remained. But even most of the diehard objectors threw in the towel a decade later, when the crater itself was identified by subsurface gravitational measurements and analysis of drill cores. The crater, known as Chicxulub after the nearby town of Puerto Chicxulub (appropriately meaning 'horn of the devil' in the Maya language), is a buried, circular, 180-km-diameter (110 miles) structure on the northern coast of Mexico's Yucatán Peninsula. Because it is covered with more recently-formed marine limestones, it had not been immediately recognizable.

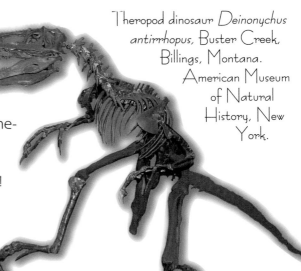

Theropod dinosaur Deinonychus antirrhopus, Buster Creek, Billings, Montana. American Museum of Natural History, New York.

    Ok, so the Cretaceous ended with a BANG! Does that explain the mass extinctions?

    Yes, and no. Clearly, such a large impact must have had a tremendous effect on the environment, although the exact nature of the disruption is highly speculative. The evidence for global forest fires is slim, but the boundary clay (which is the altered remnants of a layer of volcanic ash) strongly indicates that a tremendous amount of particulate matter was thrown up into the atmosphere. It would be remarkable indeed if this did not drastically affect the weather and the animals and plants living at the time. But how? Were sunlight to be blocked for an extended period of time, most plants would die. Large animals need more resources than small ones, and if food were scarce—either plant food or prey animals—the largest animals would be the most vulnerable to starvation. This could explain the selective loss of the large dinosaurs. Alternately, a prolonged cold spell might wipe out all the dinosaurs that weren't insulated with feathers. Small, furry mammals may have had an advantage here. Temperatures world-wide must not have dropped below freezing, however—even for a short while—or we would not have palm trees still with us. Aquatic lifestyles buffer reptiles against extremes of temperature, but crocodylians are intolerant of cold—another indication that the 'impact winter' part of the scenario could not have been too harsh. Nor could it have been less severe but of long duration—there is no evidence of polar ice forming at this time. The oceans, due to their large thermal mass, must have suffered less insult than rivers and lakes—yet it was the marine crocodiles that went extinct, and the freshwater ones that survived: exactly the opposite of what one would expect if low temperatures were a primary culprit. Why did the oceangoing mosasaurs go extinct, while crocodiles made it? Could it be that mosasaurs instead starved to death, because the ammonites all died off? It has been suggested that ammonite larvae fed on plankton (which were hard hit), while nautiloid young didn't, thus explaining the demise of the former and the survival of the latter. Are nautiloids so different from ammonites that mosasaurs couldn't eat them? If there was widespread acid rain, how did frogs and salamanders manage to survive, when their modern counterparts are extremely sensitive to exactly that? The acid rain suggestion must either be erroneous, or its effects were local rather than global. It is widely believed that pterosaurs were warm-blooded and insulated with a hairy covering. What made birds better able to survive a period of cold and/or darkness? Another enigma is the loss of the cycadeoids but not the cycads—both had starchy trunks to serve as food storehouses in the event of a prolonged, dark 'impact winter,' and most living cycads are notoriously intolerant of low temperatures. It is hard to see how the vast majority of plant species could have come through the boundary event just fine if the period of darkness lasted too terribly long; most seeds do not remain viable for more than a handful of years.

Sea turtle *Protostega gigas*, Niobrara Chalk, Kansas. Denver Museum of Nature & Science, Denver, Colorado.

Many studies indicate that dinosaur diversity and abundance were declining for the last ten or so million years of the Cretaceous (although there are a few reports that contradict this). It is highly unlikely that they died of fright waiting for the asteroid to hit! Ichthyosaurs died out considerably prior to the end of the Cretaceous; pterosaurs were declining throughout the latter half of the period. Clearly, the Chicxulub impact was not responsible for these events. Gymnosperms were progressively losing ground to angiosperms, which meant big changes in the available food resources. Plate motions were slowing down, and with this came a relatively rapid retreat of the warm, shallow, epicontinental seas. This caused a major loss of habitat for shallow-water marine creatures, and a dramatic increase in the seasonality of continental climates. (Compare the seasonal extremes central Canada endures with those experienced today in Great Britain for an idea of how powerful the moderating influence of the oceans can be.) Oxygen isotope data reveal that the oceans were gradually cooling during the last few million years of the Cretaceous—another factor which has been implicated in previous mass extinctions. This parallels a decline in global temperatures that peaked during the emplacement of the Deccan Traps basalts. The eruption of the Deccan Traps most likely contributed significantly to climatic disruption toward the end of the period. The effect of the oceanic cooling can be seen in plots of the size of marine microfossil shells across the boundary, and these changes began to occur several million years prior to the asteroid impact. The time was one of drastically changing environments, which always stresses incumbent animals to adapt quickly, or die out.

So, what we have here is a time of rapidly changing conditions occurring throughout the latest Cretaceous, punctuated at the end by a huge asteroid impact. The timing and pattern of the extinctions indicate that the most likely scenario is that of a cataclysm superimposed upon a time of higher-than-usual extinction rates caused by changing climates and food supplies. The asteroid impact was the coup de grâce, but not the sole perpetrator.

Heteromorph ammonites *Didymoceras stevensoni*, Pierre Shale, Wyoming. Denver Museum of Nature & Science, Denver, Colorado.

Where Cretaceous Rocks Are Found

Cretaceous sediments are common across North America, partly because deposition was extensive in the epicontinental seas, and partly because the closer we come to the present, the less time erosion has had to scour the rocks away. The Sierra Madre Oriental of Mexico (states of Coahuila, Nuevo León, Tamaulipas, and San Luis Potosí) has large exposures of Cretaceous rocks, which continue in a disconnected fashion throughout the Cordillera as far north as Alaska. Eastern California has good exposures as well. A wide band of sediments laid down in the Western Interior Seaway stretches from Texas and Colorado northward to southwestern Manitoba and across much of north-central Montana, southern Saskatchewan, and nearly the entire province of Alberta. There are also good outcrops in the Yukon, the North-west Territories, the North Slope of Alaska, and the Canadian Arctic Archipelago. An apron of Cretaceous rocks is exposed in an arc around the southern end of the Appalachians, from western Tennessee through Mississippi, Alabama, and Georgia, and in a discontinuous strip northward through the Carolinas and into Maryland and New Jersey.

Where Cretaceous Fossils Are Found

Inoceramid clams, Lewis Formation, Sandoval Co., New Mexico. New Mexico Museum of Natural History, Albuquerque.

In the east, most of these sediments are marine, and pro-duce few dinosaur bones with the exception of disarticulated elements that washed into the ocean. The Woodbury For-mation of New Jersey bears the distinction of having yielded the first relatively complete dinosaur skeleton known to science, that of *Hadrosaurus foulkii* uncovered in 1858. Fragmentary dinosaur bones also come from New Jersey's Marshalltown and Navesink formations. A few dinosaur bones have been found in Maryland's Arundel Clay, but mostly these eastern formations yield bivalves, shark's teeth, elements from mosasaurs, crocodiles, sea turtles, and other marine fossils. Belemnites are commonly found in the Mt. Laurel For-mation of Delaware. Cretaceous marine rocks in the southeastern U.S., including the Owl Creek and Prairie Bluff formations of Mississippi and the Navarro Group of Texas and Arkansas, frequently yield a plethora of fossil oysters. In Alabama, bones of turtles, mosasaurs, and fish may be found in the Mooreville Formation.

Sediments laid down in the Western Interior Seaway produce many fine Cre-taceous fossils. Texas' Del Rio and Walnut formations are good places to look for shark's teeth, ammonites, clams, and sea urchins. The Niobrara Chalk, exposed from western Kansas north into Saskatchewan, is one of the richest formations in North America for marine vertebrates, producing a wide variety of fish skeletons, isolated vertebrae and teeth, mosasaur and plesiosaur bones, shark's teeth, and the occasional bird or pterosaur bone. Oysters are so common in the Niobrara that some layers are composed almost entirely of their shells, and occasionally a spectacular group of crinoids is found. The Greenhorn Formation of eastern Colo-rado and western Kansas preserves a similar fauna. The Pierre Shale in Colorado, Wyoming, and South Dakota produces some beautiful ammonites and baculites; its correlate in western Colorado is the Mancos Shale, which yields shark's teeth and crinoids in addition to ammonites.

The beaches flanking the seaway preserve countless footprints: Texas' Glen Rose Limestone contains several famous dinosaur track localities, as does the Dakota Sandstone of New Mexico and Colorado. The Western Interior of North America is one of the richest dinosaur bone areas in the world. The Hell Creek, Lance, and Judith River formations of Wyoming, Montana, and Alberta have all yielded fabulous dinosaurs, including ceratopsians, hadrosaurs, and theropods. The Hell Creek also produces a wide variety of other terrestrial fossils, including bones of mammals, crocodiles, lizards, turtles, snakes, and birds. Bones of fish, amphibians, reptiles, and birds have all been found in Wyoming's Mesaverde Formation. Utah's Cedar Mountain Formation has yielded ankylosaurs, hadrosaurs, sauropods, and other dinosaurs (including the famous *Utahraptor*). The North Horn and Kaiparowits formations of the same state are rich in dinosaur bones. Also in Utah, the Blackhawk Formation preserves a swamp vegetation, including palms. Montana's Kootenai Formation yields leaves of ferns and conifers. The Meeteetse Formation of Wyoming is a gold mine for Cretaceous leaf fossils, producing palms, ferns, conifers, cycads, and herbaceous angiosperms. Arizona's Shellenberger Canyon Formation contains plant fossils and petrified wood, and the Ft. Crittenden Formation yields more of the same plus dinosaur and fish teeth, turtles, and clams.

Ammonite *Placenticeras intercalare*, Pierre Shale, South Dakota. This is an example of the gemstone "ammolite." Private collection.

Further from home, northern Alaska's Prince Creek Formation contains a plethora of hadrosaur bones, and the Naushuk Group yields angiosperm leaves, ferns, conifers, and ginkgos. The Smoking Hills Formation of Canada's Arctic produces fossils of mosasaurs, plesiosaurs, fish, and toothed birds. South of the border, many fine ammonites come from various Cretaceous formations in Chihuahua, Mexico. A hadrosaur (duckbill dinosaur) find was recently reported from Honduras.

Where Can Cretaceous Fossils Be Seen in the Field?

One of the premier Cretaceous localities on the North American continent is Alberta's Dinosaur Provincial Park near Drumheller. The Dinosaur Park Formation is chock full of the bones of these behemoths, and the overlying Bearpaw Shale produces some stunning ammonites that, when polished, glitter in an array of oranges, reds, and greens. Fractured specimens (collected on private land) are cut, polished, set in jewelry, and sold as the gemstone "ammolite." If your travels take you this far north, don't miss the nearby Royal Tyrrell Museum of Palaeontology, where many fossils collected in the park are on display.

Mosasaur *Platecarpus*, Niobrara Chalk, Kansas. Sternberg Museum of Natural History, Hays, Kansas.

The Cretaceous • 163

A thousand miles to the south, several parks showcase fine examples of dinosaur trackways. Dinosaur Ridge, near Morrison, Colorado, protects exposures of the Dakota Sandstone which are petrified Cretaceous beach sands. Sets of parallel ornithopod tracks and a few theropod prints tell of herds of dinosaurs migrating along the western shore of the Interior Seaway. Clayton Lake State Park in northeastern New Mexico exposes the same formation and more tracks of identical types of dinosaurs—perhaps even some of the same individuals, further along their trek!

Ceratopsian dinosaur Triceratops, provenance unknown. Natural History Museum of Los Angeles County, Los Angeles, California.

Grand Staircase-Escalante National Monument in southern Utah, mentioned as a place where you can hike to Jurassic dinosaur footprints, also contains outcrops of Cretaceous rocks. Many dinosaur teeth and some bones can be found in the Kaiparowits Formation there.

Dinosaur Valley State Park near Glen Rose, Texas is home to some of the first identified sauropod tracks, as well as footprints of theropods. The Glen Rose Limestone forms the creekbed, so depending on the level of the river, many of the tracks may be submerged. In the middle of the twentieth century, the American Museum of Natural History excavated a partial trackway by sandbagging the river, and transported it back to New York for exhibit.

What if you want to see some remnants of Cretaceous life *other* than dinosaurs? Check out W. M. Browning Cretaceous Fossil Park at Booneville, Mississippi. The park overlooks Twentymile Creek, where the teeth of sharks, crocodiles, and mosasaurs can be found along with fossils of clams, oysters, crinoids, and snails.

They say that "March comes in like a lion, and goes out like a lamb." Exactly the opposite can be said of the Cretaceous: its dawning was quiet, but its end was sudden and dramatic. Its encompassing Era, the Mesozoic, was marked on both ends by great catastrophes. Sadly, the magnificent, majestic dinosaurs are no longer with us, but on the other hand, whilst they ruled the Earth, they kept our ancestors relegated to the shadows. Had the big reptiles not been extinguished, it is very likely that we would not be here to marvel at their fossilized remains. The story of how our little, ratlike forebears inherited the Earth in the wake of the Chicxulub catastrophe will be the topic of the next chapter.

The Cenozoic Era

North America soon recovered from the insult it had suffered at the close of the Cretaceous. An early Cenozoic springtime in southern Saskatchewan finds plesiadapiform primates bouncing through fig trees and a condylarth mother and her young, early hoofed mammals, enjoying the flush of new forage. The forest is ablaze with rhododendrons and other flowering trees.

The catastrophic close of the Mesozoic brings us to the dawn of the Cenozoic Era ("Era of Recent Life"), in which we now live. The great saurians were gone, and our ancestors—small, shadow creatures—were set to inherit the earth.

The Cenozoic is frequently called the "Age of Mammals," though the "Age of Flowering Plants" or "Age of Birds" might be just as apt. (Strange what anthropocentrism does!) The Cenozoic is divided into two Periods, the Tertiary and the Quaternary. They are of very unequal duration, with the Tertiary spanning more than 95% of Cenozoic time.

At this point in our story, we will begin to examine the history of our continent on a finer timescale than we have thus far. Each of the previous chapters was devoted to an entire geologic Period of time; now we will break the Periods of the Cenozoic into the next-smaller division, the Epochs. There is a good reason for this, and it's not that I consider the evolution of mammals to be more important than that of invertebrate animals or "lower vertebrates" (fish, amphibians, reptiles). (Although some of my best friends are mammals, and I even married one, I've always been partial to lower vertebrates myself.) It's because the resolution of the fossil record improves significantly as we approach the recent, mainly because there has been less time for Mother Nature's destructive forces to erode the rocks away. A better fossil record means that we know the history of life in more detail.

The dawn of the Cenozoic, or start of the Tertiary, found North America recovering from the aftermath of the Chicxulub impact's destruction. The slate hadn't quite been wiped clean, but many of the previous players were gone. It was to the survivors that the future would belong.

The Tertiary Period

Two multituberculates scurry up a maple tree after a spring rain in Oregon 45 million years ago. Strange knobby-faced uintatheres browse on low-growing palms, heedless of the restless volcanoes on the horizon.

The Tertiary Period, as noted, covers the first 95% of Cenozoic time. It is comprised of five Epochs (from oldest to youngest): the Paleocene, Eocene, Oligocene, Miocene, and Pliocene. Each of these will receive a chapter of its own in turn.

The name Tertiary was first used in 1760 by Giovanni Arduino for the third section of his tripartite division of the montane rocks in northern Italy (after the 'primary' and 'secondary,' known to us today as the Paleozoic and Mesozoic eras). In 1828, Charles Lyell proposed a set of epochal subdivisions for the Tertiary. The suffix 'cene' means 'recent,' and Lyell used it, with modifiers, in the names of all of the Tertiary epochs. A version of his naming scheme is still in use today; however, the original system only used three epochs—the Eocene, the Miocene, and the Pliocene. Later workers added the Paleocene and Oligocene epochs to the Tertiary series.

The Tertiary was a time of fine-tuning of the geography of North America, during which the present topography, coastlines, and drainage patterns were set. Tectonic motion shifted from the largely northward path our continent had taken since the late Paleozoic to primarily westward drift. We have seen that by the latest Cretaceous, the geography of North America (and the other continents) was coming to quite resemble the world map of today. There were, however, minor differences: the Atlantic Ocean continued to open during the Tertiary, pushing our continent further from Europe. Most of the width of the Atlantic has been achieved during Tertiary time. Greenland, a part of North America since the Precambrian, initially stayed with Scandinavia but by around 45 mya it broke away, becoming a microcontinent of its own. This is important, because during the early Tertiary, the seaways both to the east and west of Greenland were narrow enough for plants and animals to cross, maintaining strong floral and faunal ties across all the northern continents. Later in the Tertiary, another land route opened between Alaska and Siberia, allowing many more migrations between North America and Eurasia.

The Tertiary story is a tale of grand success for the mammals that had survived the K/T event. We will see how our small, rodent-like forebears gave rise to the stunning array of mammals that populate today's world, as well as many bizarre extinct forms. To no lesser extent it is a tale of success for the birds as well, but they will play a smaller role in our narrative for one simple reason: for the most part, their bones are so delicate that they do not fossilize well, hence we know less about them than we do of larger creatures. Early mammals are to some extent in the same boat, but they have left us a rich record of fossil teeth that the birds have not—giving us a glimpse into size and diet, at least.

Where Tertiary Rocks Are Found

Up until now, we have noted where in North America one can find outcrops of rocks from each specific time period. Most geologic maps, however, do not differentiate rock units at the Epoch scale, grouping them at best into Lower and Upper Tertiary. We'll still continue to denote specific fossil-bearing formations, but of necessity give only an overview of Tertiary rock exposures.

North America has an extensive Tertiary fossil record. We owe the formation of many of these rocks to outwash from the Rocky Mountains and ashfalls from the Cascade volcanoes. Early Tertiary rocks are largely absent from Canada, save for an apron of sediments in southwest Alberta adjacent to the Rocky Mountain front. Most of western Mexico's Early Tertiary rocks are volcanic, but the Gulf states of Tamaulipas, Veracruz, Campeche, and Yucatán, and on the Caribbean side, Quintana Roo have a good Early Tertiary sedimentary record. The United States has the best exposures of Early Tertiary rocks of any North American country. A large swath of such rocks cuts diagonally across Texas, northern Louisiana, and Arkansas in a northeast-southwest direction. Passing east of the Mississippi River valley (where these rocks have been worn away), Early Tertiary sediments form a broad apron southwest, south, and southeast of the Appalachians, wrapping in a continuous arc from western Tennessee south through the Gulf states and north into the Carolinas. The northern Gulf Coast of Florida also has some Early Tertiary rocks, and there are disconnected exposures in Oregon and Washington state. The midcontinent and northeastern states generally lack such sediments, but they are encountered again throughout the eastern Rocky Mountains and in eastern Montana and the western Dakotas.

Late Tertiary exposures are even more extensive, as befits the lesser amount of time there has been for their removal by erosion. Again, Canada has no Late Tertiary record to speak of—we have glacial scouring to thank for that—but the central mountains of Mexico (states of Chihuahua, Durango, Zacatecas, Guanajuato, Hidalgo, and Puebla) have good exposures, as do several Central American countries (Nicaragua, Costa Rica, and Panama). Small outcrops of Late Tertiary sediments are scattered throughout the western mountains of the United States, but the most extensive exposures are east of the Rockies from New Mexico to Wyoming, and parallel to the Gulf and Atlantic coasts. An almost continuous band of Late Tertiary sediments stretches from New Jersey south to Georgia and west to the Texas-Mexico border. Most of the western half of Florida is also underlain by Late Tertiary rocks.

In the interior of the continent these rocks are mainly terrestrial, as the great epicontinental seaways withdrew from the continent in the waning days of the Cretaceous. But without waters locked up in polar ice until mid-Tertiary times, sea levels were still significantly higher than today, and marginal areas of North America hosted more extensive continental shelves than they do now. Thus states bordering the Gulf of Mexico and Atlantic do hold some marine formations.

Enough generalities. We left North America, at least the western Gulf Coast and the southern Western Interior, leveled by the end-Cretaceous meteorite blast. The Paleocene found our continent struggling to recover from the devastation—and it is to this story that we will now turn.

Chapter Twelve
The Paleocene

The last incursion of the waves into the interior of our continent was the Paleocene Cannonball Sea. A slough in North Dakota provides refreshment for a pantodont, while a ringtailed multituberculate scampers up a sycamore tree. Bald cypresses line the waterway; magnolias and ancestors of cherry trees choose the higher ground.

Paleocene North America

The Paleocene, the first epoch of the Tertiary Period and of the Cenozoic Era, found our continent recovering from perhaps the greatest insult it has suffered in its known history. Immediately after the terminal Cretaceous meteorite impact, the Gulf Coast region of North America was a wasteland extending as much as 1000 miles (1600 km) inland. We have already mentioned, in the last chapter, the fern spore 'spike' immediately above the K/T boundary clay, which is interpreted as revealing initial recolonization by ferns following the devastation. The further we get from the Chicxulub crater, the less pronounced this fern spike is—it is stratigraphically thinner (and of shorter duration) in North Dakota than in New Mexico, smaller still in Japan, and nonexistent in New Zealand and Antarctica. This makes sense, that the lands closest to the impact would be hardest hit, so to speak. Yet within a few thousand to half a million years after the K/T event, vegetation had rebounded and mammals had begun to diversify into an astonishing array of shapes and sizes—now that they were no longer kept in check by the dinosaurs.

Definition and Nomenclature

The name Paleocene means "early dawn of the recent." The term Paleocene was first used by W. P. Schimper in 1874 for the distinctive European flora of the earliest part of Lyell's Eocene. The Cretaceous-Tertiary boundary (start of the Paleocene) was defined in 1991 as the iridium geochemical anomaly at the base of

the boundary clay, El Kef, Tunisia, with a date of 65 million years ago. The Paleocene-Eocene boundary, or close of the Paleocene, was chosen in 2003 as the base of a negative carbon-isotope excursion at the Dababiya Section near Luxor, Egypt, dated to 56 mya. Thus the Paleocene may be considered to span some 9 million years.

Paleogeography and Paleoclimate

As the Atlantic widened, North America drifted steadily westward, overriding the Pacific seafloor. This caused the leading (western) edge of the North American plate to buckle, thrusting up the Sierra Nevada and Coast Ranges, and even raising the Rockies as much as a thousand miles inland. This episode of mountain-building, known as the Laramide Orogeny, had begun in the latest Cretaceous and continued throughout the Paleocene. At that time, however, the Rockies were not nearly as high as they are now—perhaps only a little taller than the modern Appalachians. Renewed uplift later in the Tertiary would raise them to their present lofty heights.

Walnut-like leaf Polyptera manningii, with unidentified pods. Ft. Union Formation, Rock Springs, Wyoming. Private collection.

The great midcontinent seaway breathed its final gasp during the mid-Paleocene: the last marine transgression of any significance had the waves of what is known as the Cannonball Sea lapping shores in North Dakota and Montana for a short while, around 60 mya. In short order, the waters retreated south to the Gulf of Mexico, although into late Tertiary times the Mississippi Embayment, Gulf Coastal Plain, and the Atlantic Coastal Plain were frequently inundated, as were the coastal areas of the Pacific margin. Much of Oregon and Washington were under water, as the region that would later host the Columbia River valley was a large embayment of the Pacific. Florida and the Bahamas consisted mainly of offshore carbonate platforms (coral reefs).

Walnut-like nut Polyptera manningii, Ft. Union Formation, Rock Springs, Wyoming. Private collection.

Paleocene climates were warm by today's standards, but considerably cooler and moister than the Cretaceous had been. It has been speculated that this was due to the effects of atmospheric dust lifted aloft by the Chicxulub impact, which may have taken hundreds of thousands (or even millions) of years to totally clear. Nevertheless, Paleocene climates still lacked the steep equator-to-pole temperature gradients that characterize the modern world; the mean annual mid-Paleocene temperature at 45°N—approximately the latitude of the Straits of Mackinac between lakes Michigan and Huron—was 15°C (59°F), compared with 6°C (43°F) today. The oceans remained warm, including the deep waters, although estimates of exact temperatures vary widely.

There was a slow, modest decline in global temperatures (to about 10°C at the abovementioned latitude) during the latter half of the Paleocene, but this was suddenly broken by a dramatic rise of short duration at the Paleocene-Eocene boundary. This temperature spike—which resulted in an increase in sea surface temperatures of between 5° and 8°C (9° to 14°F) within a few thousand years—is called the Paleocene-Eocene Thermal Maximum or PETM. Something unusual must have happened for such a dramatic warming to have happened so suddenly. Although the cause is uncertain, one viable theory is that it was the release of massive amounts of seafloor methane hydrates—which would have caused a runaway greenhouse effect—such as has been implicated in the end-Permian extinction event. Carbon isotope data supports this theory, but why associated faunal turnover was nowhere near as drastic this time remains a mystery.

Life in the Paleocene

Paleocene sea life was scarcely distinguishable from what we see today, with the important caveat that marine mammals were absent. Teleost fish dominated the vertebrate realm, and with the great marine reptiles gone, sharks reigned as top predators. The ubiquitous ammonites of the Mesozoic were replaced by soft-bodied squid. Although a few relict Paleozoic groups, such as the brachiopods and crinoids, hung on by a thread, the invertebrate fauna of Paleocene times was characterized by an abundance of snails, clams, and (in warm waters) hexacorals, much as it is today. Species were different, of course, but in general aspect, a Paleocene seafloor would look much like a modern one.

Continuing what they had begun during the Cretaceous, angiosperms or flowering plants made consistent inroads into the terrestrial flora. The widespread use of insect pollinators was likely one factor in their success. Conifers still held an advantage at high latitudes, where their ability to photosynthesize year-round gave them an edge. Tropical and midlatitude floras became overwhelmingly dominated by flowering plants, including both deciduous and evergreen broadleaf trees. The interior of the continent received abundant rainfall, because the western mountains were not yet high enough to block moist winds from the Pacific. Extensive basins associated with the rising Rockies hosted widespread swamplands from Colorado to the Arctic. The northern forests were dominated by water-loving deciduous conifers like *Metasequoia*, with an understory of ferns,

Sycamore *Platanus raynoldsii*, Denver Formation, Colorado. Denver Museum of Nature & Science, Denver, Colorado.

horsetails, and ginger-like herbs; from Wyoming southward, the conifers were replaced by broadleaf angiosperms and palms.

Amongst terrestrial animals, the meek truly had inherited the Earth. For the first two-thirds of mammalian history, they had stayed very small, eking out a living in the shadows and staying out of the way of the great saurians. Now all the econiches available to larger animals were vacant, save for a few crocodiles. In short order the surviving mammals had spawned a wide variety of new forms. Within the first quarter-million years of the Paleocene, mammalian diversity in western North America increased from 20 to 45 genera; 70 genera were present 2 million years later. By the end of the epoch, 120 families of mammals are known from the fossil record—a comparable diversity to that of the modern day. This was the most important radiation in mammalian history. In addition to primitive representatives of several modern mammalian groups, the Paleocene also saw the evolution of many bizarre groups of mammals that are no longer with us. North America has the best fossil record of Paleocene mammals of any continent, including postcranial skeletons of some critters known elsewhere only by their teeth.

We've already met the multituberculates, those small rodent-like mammals with the mountain-bike-tire molars. They skated across the K/T boundary relatively unscathed, and were common animals in Paleocene ecosystems, reaching their peak of diversity at this time. However, their days were numbered, as late in the epoch the first true rodents appear in the fossil record; competing for the same econiches and resources, the rodents ultimately won out.

The Order Insectivora, which includes modern shrews, moles, and hedgehogs, is considered to be the most primitive living placental order. These insectivores and omnivores with small brains may be similar to many Mesozoic mammals and representative of the stock from which many Paleocene groups arose. They were abundant and diverse in their own right; most were mouse- to opossum-sized, and some lived in trees. (One known from Alberta even had grooved teeth demonstrating that it was venomous!) But they didn't have the trees to themselves—North America during the earliest Tertiary was also home to early primates (or primate relatives) called plesiadapiforms. These were squirrel-like or lemur-like animals that fed on the abundant fruits that the flowering trees provided, as well as perhaps nectar and gum.

Palm Sabalites sp., Denver Formation, Colorado. Denver Museum of Nature & Science, Denver, Colorado.

Perhaps the quintessential Paleocene mammal group is the pantodonts, dog- to bear- sized animals with a stocky build, clawed feet, an elongated face, and a long tail. They were rather generic mammals that probably browsed in forests and along rivers. Strange animals called taeniodonts and tillodonts were the rooters and diggers of their day. These medium-sized, mostly heavily-built animals had long claws and large tusks which they used to dig up underground tubers.

Phenacodont condylarth *Tetraclaenodon* sp., Nacimiento Formation, New Mexico. Private collection.

Sometimes called a "wastebasket category" because of the uncertain evolutionary relationships of its members, the condylarths were the first mammals to evolve hoofs on their toes. They were also some of the first mammals to specialize in leaf-eating. The earliest condylarths are rat-sized, but by the end of the Paleocene, many were as large as sheep and a few were even bear-sized. The ungulates (hoofed mammals) of today derive from within this group. Condylarths were extremely common; they account for more than half the mammal specimens in the latest Paleocene fossil record of North America.

At the dawn of the Paleocene, there were essentially no large mammalian predators. This void was soon filled by several types of animals, including some of the condylarths. The first mammals assigned to the Order Carnivora (which today includes cats, dogs, weasels, and bears) appear in the North American Paleocene, but the most common predatory mammals of the time belong to a group called the creodonts, a separate lineage. Some resembled cats in general form and may have been able to climb trees, whereas others were heavily-built and hyena-like in general appearance. Their powerful jaws suggest that like hyenas, they may have been able to crack bones. Creodonts are distinguished from true carnivores by their usual possession of two pairs of carnassial (shearing, scissor-like) teeth, vs. only one pair in true carnivores; which specific teeth take on the carnassial form also differs between the two groups. Early Tertiary predators like the creodonts were not built for fast running, and probably ambushed their victims. Although initially more diverse and numerous than the Carnivora, the creodonts became extinct by the mid-Tertiary, possibly because of competition from true carnivores.

A Paleocene Vignette

Imagine, if you will, a Paleocene forest somewhere in western North America. Up in a stand of fig trees, a family group of primate-like mammals, called *Purgatorius*, noisily forages for fruits like a troop of monkeys: but they have longer bodies and faces than monkeys, and claws on their prehensile toes—somewhat like a three-way cross between a pygmy marmoset, a rat, and a squirrel. Primates seem to have originated in North America, only

Dove tree *Davidia antiqua*, Ft. Union Formation, Rock Springs, Wyoming. Private collection.

later spreading across Greenland into Eurasia before becoming extinct here. Down below, another odd beast, a tillodont—looking something like a peccary-sized blend of a bear and a capybara, with large rodent-like incisors and long claws—roots about in the underbrush. There are no very large or long-legged animals here—those are not useful adaptations in a dense forest. An owl dozes lazily on the branch of a cypress tree, looking recognizably modern, his white plumage almost blinding against the forest's deep. A pantodont, a nondescript mammal the size of a small black bear that might be a slender, clawed sheep-tapir cross with a long tail, browses on the tender leaves of a willow shrub.

Where Paleocene Fossils Are Found

Given the extensive North American Tertiary record, it is not surprising that there are many fossiliferous rock units to be found here. Some of the earliest small, generalized Paleocene mammals come from the upper portion of Colorado's Denver Formation, which straddles the Cretaceous-Tertiary boundary. This formation also yields a diverse fossil flora of angiosperms, cycads, ferns, conifers, and seeds. The Nacimiento Formation of southern Colorado and New Mexico is a classic unit for Paleocene mammals, and also contains plants, fish, turtles, and crocodiles. The Ft. Union Formation of Wyoming, Montana, and North Dakota produces a stunning variety of fossil leaves, mostly of angiosperms. Hundreds of individual fossil-leaf localities are known from this one formation alone. The Cannonball Formation of North Dakota was laid down in and near the shores of the Cannonball Sea and contains fossils of crocodiles, fish, clams, snails, birds, and some petrified wood. The Sentinel Butte Formation of the same state has produced fossilized mammal teeth, including multituberculates and plesiadapiforms, as well as beautiful leaves from ginkgo, birch, and other trees. Fish and some condylarths are found in Wyoming's Ferris Formation, and mammal teeth come from that state's Hanna Formation as well.

Ginkgo leaf *Ginkgo adiantoides*, Sentinel Butte Formation, Almont, North Dakota. Private collection.

Paleocene snails, oysters, and crocodile and shark's teeth come from Virginia and Maryland's Aquia Formation, as well as some very nice pine cones. South Carolina's Williamsburg Formation produces some plant fossils as well as an assortment of molluscs. The Naborton Formation of Louisiana yields abundant fossil leaves, usually found in clay lenses associated with coal deposits. California's Silverado Formation produces a flora very similar to that from the Ft. Union, but with fewer conifers and more birch and witch hazel. Clams and snails are commonly found in the Wills Point Formation of Texas. Illinois' Clayton Formation produces clams, snails, bryozoans, crabs, lobsters, and shark and crocodile teeth. The most diverse Paleocene freshwater fish fauna known from North America comes from Alberta's Paskapoo Formation. The Chickaloon and Tolstoi formations of southern Alaska yield a mixed evergreen/deciduous flora similar to that growing in north Florida today. Other fossil forests are known from Alaska's

North Slope, Greenland, and Ellesmere Island in the Canadian Arctic, attesting to the warmth of high latitudes during the Paleocene.

Although fieldwork south of the border has been sparse, abundant brachiopods, snails, echinoids, corals, and bryozoans have been reported from fossiliferous lenses in the Potrerillos Formation in the La Popa Basin of northeast Mexico.

Where Can Paleocene Fossils Be Seen in the Field?

The remains of a diverse flora and fauna representing a subtropical swampy forest can be visited at the Wannagan Creek Site in Theodore Roosevelt National Park near Medora, North Dakota. This locality, with exposures of the Bullion Creek and Sentinel Butte formations, has yielded many Paleocene mammals including multituberculates, squirrel-like primates, and condylarths; reptile fossils are common, with turtles, varanid lizards, and alligators all represented. Fish, clams, and snails are abundant. Many trees with modern analogues are present, including palm, bald cypress, dogwood, fig, elm, magnolia, and sassafras. The most common fossil in the park is petrified wood, and in the Petrified Forest Plateau area there are even some stumps standing in life position.

Walnut-like leaf Polyptera manningii, Ft. Union Formation, Rock Springs, Wyoming. Private collection.

The Paleocene saw our continent recover from the Chicxulub event, and our small, furry forbears diversify to become the dominant animals in terrestrial ecosystems. Its close was marked by a dramatic warming that would usher in the Eocene, the last great time of tropicality for North America.

Chapter Thirteen
The Eocene

By the Eocene, modern coral reefs were established and whales were top predators in the seas. The primitive whale *Zygorhiza* cruises the shallows of Louisiana above scallops, lobsters, sea grasses, a horseshoe crab, a sea turtle, and fish of every stripe and color.

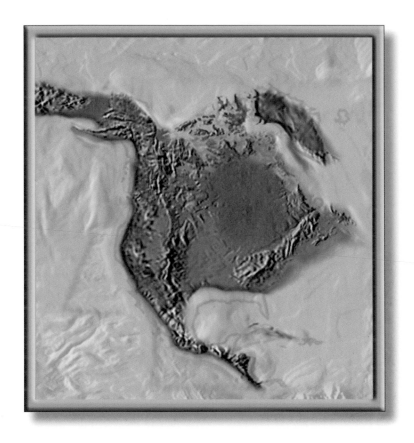

Eocene North America

North America cloaked from shore to shore in tropical and subtropical rainforests: that was the Eocene world. The PETM temperature spike—probably caused by an infusion of seafloor methane into the atmosphere—atop the general global warmth of the bygone Mesozoic Era ushered in the last great age of tropicality that our continent would see. Beginning in the mid-Eocene, a slow, inexorable decline in worldwide temperatures set in, eventually leading to the advance of the great ice sheets.

Definition and Nomenclature

The Eocene—its name meaning "dawn of the recent"—is the second Epoch of the Tertiary Period. The Paleocene-Eocene boundary, or beginning of the Eocene, was set in 2003 at the base of a negative carbon isotope excursion in the Dababiya Section near Luxor, Egypt, with a date of 56 million years ago. Its ending, at the Eocene-Oligocene boundary, was defined in 1992 at the extinction horizon of the planktonic foraminifer *Hantkenina* at the base of the marl bed 19 m above the base of the Massignano quarry at Ancona, Italy. This event is currently dated at 34 mya, giving the Eocene a duration of approximately 22 million years.

Paleogeography and Paleoclimate

By this time, the world map of today was recognizable, and the geography of North America differed from its modern counterpart only in detail. The general out-

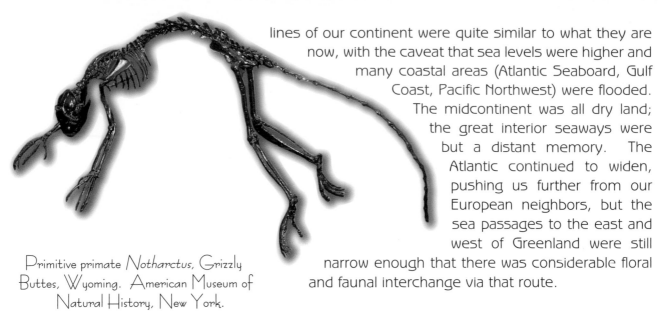

lines of our continent were quite similar to what they are now, with the caveat that sea levels were higher and many coastal areas (Atlantic Seaboard, Gulf Coast, Pacific Northwest) were flooded. The midcontinent was all dry land; the great interior seaways were but a distant memory. The Atlantic continued to widen, pushing us further from our European neighbors, but the sea passages to the east and west of Greenland were still narrow enough that there was considerable floral and faunal interchange via that route.

Primitive primate *Notharctus*, Grizzly Buttes, Wyoming. American Museum of Natural History, New York.

With today's seven continents distinct and spread across the globe, moist maritime air masses had access to most interior areas, and rainfall was generally high. Rainforests shrouded North America from the Atlantic to the Pacific, and from the Caribbean to the Arctic. Subtropical zones extended to as much as 60°N, and even Alaska, Greenland, and the Canadian Arctic isles enjoyed a temperate, largely frost-free climate. Mangrove thickets lined the Gulf of Mexico coast as far north as Texas. The Rockies continued to rise, but did not yet cast an appreciable rain shadow over the continental interior. Intermontane basins in the region where Wyoming, Utah, and Colorado now join hosted vast freshwater lakes teeming with aquatic life, much like—though significantly warmer than—the Great Lakes of today.

But plate tectonics is restless, and this tropical paradise could not last. A world away, South America and Australia were wrestling free of Antarctica, and the latter was settling in over the South Pole. With the opening of the Drake Passage and the Southern Ocean, what is called the Antarctic Circumpolar Current was born (although, at first, it only flowed at fairly shallow depths, having to still traverse the continental shelves and Macquarie Ridge). This stream flows from west to east around the perimeter of Antarctica, mixing very little with waters to the north. The lack of commingling deprives the circumpolar waters of an influx of equatorial heat, and as a result, allows them to become very cold. By the mid-Eocene, a significant latitudinal temperature gradient became established in the Southern Hemisphere. This ultimately—as the Tertiary progressed—led to glaciation on Antarctica, which had significant ramifications for climate the world over. A cooling and concomitant drying trend set in that eventually culminated in the Quaternary Ice Ages. This would take some 50 million years, however, and by modern standards, the Late Eocene was still quite warm.

We've discussed in detail heretofore the asteroid impact that terminated the Cretaceous. As stands to reason, impacts of a lesser magnitude also happen from time to time. Around 35 mya, a bolide some 3-5 km (2-3 miles) in diameter struck our eastern continental shelf in the area that is now Chesapeake Bay. It created a slurry of water and fragmented rock in the crater proper which has ultimately determined the location of the Bay, although the present coastline wasn't actually sculpted until some tens of thousands of years ago. The fractured rock in the crater has, however, become more susceptible to settling and erosion than the intact bedrock surrounding it, allowing more recent rivers to downcut preferentially in this area and thus determine the direction of drainage in the region.

This astrobleme is a bit too old to be the sole culprit in the spate of extinctions that occurred at the end of the Eocene, but it may have been a contributing factor. At twice the size of Rhode Island and as deep as the Grand Canyon, the buried crater indicates an impact that is certainly big enough to have caused significant destruction in a local area and possibly to have had global effects as well. The pattern of species loss corresponds better with a scenario of gradual climatic change than it does with a singular catastrophe, but as with other extinction episodes, the factors are likely to have been many.

Life in the Eocene

We have noted in the last chapter that by now, marine faunas were bivalve- and gastropod-dominated with extensive hexacoral reefs, much as they are today. To a nonspecialist, an Eocene reef would be scarcely distinguishable from a modern one. Sea urchins, crabs, lobsters, and squid were common inhabitants; teleost fish and sharks plied the open waters. Foraminifera were the most abundant plankton; one species, *Nummulites*, grew as large as a quarter-dollar, and its fossils may be seen embedded in the limestone used to construct the Great Pyramid at Gizeh.

The first marine mammals to appear in the early Tertiary seas were the primitive whales, or archaeocetes, which made their debut some 40 or so million years ago. In North America, fossils of these primitive whales come primarily from the Gulf Coast states of Louisiana, Mississippi, and Alabama. As with the great marine reptiles of the Mesozoic, these oceangoing predators had terrestrial ancestors. For many years it was thought that early whales could trace

Palm fronds and fish, Green River Formation, Wyoming. Las Vegas Museum of Natural History, Las Vegas, Nevada.

their lineage back to a primitive group of placental mammals called mesonychids, but recent fossil and molecular genetic evidence points to their derivation from hoofed mammals called artiodactyls (see later). Archaeocetes still possessed small limbs, belying their landlubbing roots.

The earliest sea cows (manatees, dugongs) come from nearshore Eocene rocks in the Caribbean and Tethys regions. Their distribution is closely tied to that of sea grasses (the only flowering plants to successfully invade the marine realm), although some later forms ranged far north along the Pacific coast of North America, and probably fed on the kelp forests which became established during the later Tertiary.

Black fly, Green River Formation, Douglas Pass, Colorado. Entire fossil measures about 3 mm long; even the eye and wing can be seen. Author's collection.

The vast tropical rainforests were comprised of a wide variety of broadleaf evergreen trees, along with palms and deciduous conifers like the dawn redwood *Metasequoia*. Modern types of moths and butterflies flitted in and out of these forests. Ferns, sedges, cattails, and water lilies lined the lakeshores and riverbanks, and flowers were everywhere, including members of the rose family. As temperatures peaked in the mid-Eocene and thereafter began to decline, deciduous broadleaf trees—which are better able to cope with seasonal variations in temperature and water supply—slowly replaced many of the tropical evergreen trees. Pine, beech, oak, willow, sycamore, aspen, sassafras, walnut, birch, maple, and white cedar are species commonly seen in Late Eocene floras. Grasses, which had first appeared in the Late Cretaceous fossil record but are known from North America only since the Eocene, still were a minor element of the vegetation. Peat accumulated in the subtropical forests in the Canadian Arctic, attesting to the high level of rainfall. The trees here were mostly conifers, but also included ginkgo, birch, walnut, sycamore, and oak.

By the dawn of the Eocene, almost all of the modern mammalian orders are represented in the fossil record, and most of the odd Paleocene experiments had died out or were on the wane. Rodents, primates, and carnivorans were already here; now they were joined by a wide variety of ungulates (hoofed mammals), which are descended from the generalized mammals called condylarths which we met in the last chapter. There are two main lineages of ungulates: the artiodactyls (even-toed or cloven-hoofed mammals) and the perissodactyls (odd-toed hoofed mammals). The difference between them is that the weight-bearing axis of the foot in perissodactyls passes through the middle (third) toe, whereas in artiodactyls, the weight is borne equally by the third and fourth toes. In both groups, non-weight-bearing lateral toes may be reduced or lost. It is unclear exactly where the artiodactyl and perissodactyl lineages originated, although it was somewhere in the holarctic realm (North America-Europe-Asia). Faunas from this timeframe across the boreal regions share many elements, and migrations back and forth via Greenland were common. Perissodactyls, which were the more numerous of the two groups in the beginning,

Archaeocete whale *Basilosaurus*, provenance unknown. University of Michigan Exhibit Museum of Natural History, Ann Arbor.

consist of horses, rhinos, tapirs, and the extinct brontotheres and chalicotheres (which we shall introduce in a moment). The artiodactyl group contains modern cattle, bison, sheep, pronghorns, deer, goats, swine, camels, antelope, and some extinct groups such as the oreodonts. They are the most numerous ungulates today by far, both in North America and elsewhere, having increased in diversity starting in the Late Eocene, at the same time that perissodactyls declined—possibly due to competition, although this is far from certain. Artiodactyl genera today outnumber perissodactyls by 10:1. One explanation for this trend may be differences in their digestive systems. Both groups are mainly herbivorous, and require copious intake of low-nutrient foliage. Bacteria in the gut are needed to digest cellulose. Artiodactyls, with their multiple 'stomachs' and foregut fermentation, are more efficient at extracting the nutrients from the vegetation they eat than are perissodactyls, with their hindgut fermentation process. This allows artiodactyls to get by on a lesser quantity of food—important when it is scarce—and may be the secret to their success.

Let's look at the perissodactyls first, because they were initially the most diverse. Early in the epoch, great knobby-skulled beasts called uintatheres (relics of a Paleocene stock) were the largest mammals in North America (some up to four tons). By the Late Eocene, a perissodactyl group called the brontotheres (= titanotheres)—which probably originated here—had supplanted the uintatheres as megaherbivores. Brontotheres looked something like huge rhinos with a slingshot on their noses instead of a horn, and could stand two meters tall at the shoulder. They went extinct in North America at the end of the Eocene, although some survived into the Oligocene in Asia.

Sycamore leaves *Macginitiea* sp., Green River Formation, Wyoming. Denver Museum of Nature & Science, Denver, Colorado.

Rhinos immigrated to our continent from Eurasia in the mid-Eocene, but attained their greatest success in mid-Tertiary times. A rhino-like group, the hyracodonts or "running rhinos," appeared around the same time. These long-legged, sheep-sized perissodactyls were some of the most cursorial (fastest-running) animals in North America at the time, with limb proportions similar to those of contemporaneous horses. Amynodonts looked like rhinos with horsey faces; some were semiaquatic riverine animals, while others were forest browsers.

The birthplace of equids (horses) is unclear, but their radiation and success is essentially North American. The "dawn horse" *Hyracotherium* (eohippus) was a small forest browser present in the Early Eocene of our continent's fossil record. The trend of evolution in the horse lineage, from the Late Eocene onward, has been to increasing size, higher-crowned teeth, and faster running ability—but most of this story is told by later Tertiary fossils. Horses were quintessentially North American until their extinction here during the Ice Ages; they were reintroduced by the Spaniards in the 15th century.

Tapirs, today considered "living fossils," also originated here in the Early Eocene. As tropical-forest animals, they enjoyed great success in North America throughout the epoch, but declined in abundance with the retreat of their forest habitat. They, too, inhabited North America until quite recently—perhaps still being here when humans arrived—but presently survive only in South America and southeast Asia.

It might seem odd that the chalicotheres—which looked something like a short-necked giraffe with the head of a horse—would be considered perissodactyls, since they possessed clawed feet, but chalicotheres were definitely odd beasts. Primarily a Eurasian group, they immigrated to North America in the Eocene and enjoyed modest success here. Their front limbs were longer than their hind ones, giving them a sloped-back, giraffey look, and they probably used their powerful forelimbs to reach browse high in the trees.

Early horse *Hyracotherium vasacciense*, Wind River Valley, Fremont Co., Wyoming. American Museum of Natural History, New York.

Artiodactyls, as mentioned, got off to a slower start. One very successful native North American group is the oreodonts, which were generalized, sheep-sized mixed browsers. They became increasingly numerous as the climate cooled and became drier and the forests thinned, being found in vast herds in some formations in the Great Plains states.

Peccaries, too, originated here in the middle of the epoch. Not known for their agreeable dispositions, their larger relatives, the entelodonts, may have been even more contentious. These "terminator pigs" stood 2 m (6 ft) high at the shoulder and were adapted for an omnivorous diet that included a significant fraction of meat. No doubt the oreodonts had much to fear.

Although today we might think of camels as Old World beasts, the lineage traces its roots back to our Eocene homeland. The llamas and alpacas of the Andes are also emigrants from up north, though they did not enter South America until a few million years ago. But for most of the Tertiary, camelids of one sort or another were common elements of the North American fauna.

In the last chapter, we discussed the creodonts and their competition with the true carnivores for the predatory econiches. The Eocene saw the former group continue to decline as the Carnivora enjoyed increasing success. Our continent was the birthplace of the canid lineage (which today contains dogs, foxes, and wolves); this group was exclusively North American for most of its history, only spreading to Eurasia in the Late Miocene or Early Pliocene. Ursids (bears), on the other hand, were immigrants to our homeland from Europe in the Late Eocene. Catlike nimravids—some sabertoothed—appear simultaneously in the Late Eocene fossil record of Europe and North America, obscuring their origins. And strange creatures called amphicyonids or "bear-dogs" (which did indeed look like a cross between a bear and a stocky dog) completed the predatory menagerie.

Bat *Icaronycteris index* (cast), Green River Formation, Wyoming. Denver Museum of Nature & Science, Denver, Colorado.

The multituberculates, which had been so successful since the Mesozoic, were in decline, presumably from competition with the rodents. Sadly, their days were numbered. Springing fully-formed and flight-capable from Eocene rocks in Wyoming is the earliest-known bat, *Icaronycteris index*. Its fully modern suite of characters clearly obscures a long evolutionary history undocumented by the fossil record.

What were the birds doing all this time? Well, they had lost their Mesozoic teeth and long bony tails, and were diversifying into the array of ecotypes that we have in the modern world. Vultures, pelicans, and quail all appeared during the Eocene. But some were rediscovering their dinosaurian roots: this timeframe also saw the appearance of large flightless birds called diatrymas—fearsome six-foot-tall predators that looked like an amalgamation of a toucan, an ostrich, and a *T. rex*. These died out in North

Shrimp *Procambarus primaevus*, Green River Formation, Wyoming. Private collection.

America by the end of the epoch, probably due to competition with more-efficient mammalian carnivores.

Where Eocene Fossils Are Found

The state of Wyoming is a goldmine for Eocene fossils, with many productive formations throughout the state. The Green River Formation of southwestern Wyoming, northeastern Utah, and northwestern Colorado is a series of lacustrine (lake) sediments most famous for its abundant fossil fish, but which, in various layers, also yields crocodiles, turtles, insects, birds, many types of flowering plants (including magnificent fan palms), and the bat *Icaronycteris*. The Green River is the remnant of sediments laid down in lakes Uinta, Gosiute, and Fossil—those Eocene "Great Lakes" we mentioned earlier. Preservation is so exceptional in the fine-grained shales that details of insect wings, setae, and eyes can be seen; the only other fossilization method to preserve better detail is whole-body entombment in amber. Wyoming's Willwood Formation yields a plethora of fossil leaves. Sediments of the Wind River Formation produce angiosperm wood and 70+ types of leaves in addition to a spectacular mammalian fauna that includes creodonts, early horses, primates, true carnivores, rodents, insectivores, and marsupials. More primates along with turtles and crocodiles are to be found in the state's Bridger Formation, and the Washakie Formation yields many mammals as well. Elsewhere in the Rocky Mountain region, delicate flowers are preserved in Washington's Klondike Mountain Formation, along with more than 200 species of leaves, including cherries, apples, hawthorns, and elms. The Florissant Formation of central Colorado preserves some of the best fossil insects in the world, as well as a wide variety of leaves including those of white cedar, beech, maple, oak, willow, sycamore, sassafras, walnut, sumac, soapberry, birch, and cattails. A small Eocene locality, British Columbia's Princeton Chert (in the Allenby Formation), is world-famous for its 3D preservation of a wide variety of plants—including pines, *Metasequoia*, ferns, palms, sedges, and water lilies—and also contains some bones of turtles, fish, and mammals.

The Moodys Branch Formation and Yazoo Clay of Alabama, Mississippi, and Louisiana contain the fossil bones of primitive archaeocete whales, including the monstrous "sea serpent"-like *Basilosaurus* and the more modest-sized *Zygorhiza*. A wide variety of shark and ray teeth, as well as the delicate bones of other fish, birds, and small reptiles, come from the Potapaco sands of the Nanjemoy Formation in Virginia.

Crystallized "Blue Forest" petrified wood, Green River Formation, Fontenelle, Wyoming. Private collection.

Where Can Eocene Fossils Be Seen in the Field?

Numerous Eocene fossil parks are scattered throughout the Rocky Mountain region of the U.S. and Canada. Some of these contain vertebrate fossils, but they are few and far between in most places—as a generality, most of these parks preserve tree and leaf fossils.

Starting in the north, Driftwood Canyon Provincial Park, near Smithers in northwest British Columbia, was erected to preserve the 50-million-year-old fossil beds on the east side of Driftwood Creek. Here can be found leaf fossils (including poplar, cranberry, and alder), fish (salmon, trout), birds, and insects. Visitors are encouraged to look for and photograph fossils, but none may be removed from the park. If you're interested in collecting Eocene material in this neck of the woods, go instead to the privately-owned McAbee Fossil Beds near Cache Creek, where a similar flora and fauna is preserved and you can sign up for guided, keep-what-you-find day digs.

Moving stateside, the Stonerose Interpretive Center and Fossil Site at Republic, Washington exposes what has been called "North America's most diverse Eocene flora" in the Klondike Mountain Formation. As well as the fossil flowers (*Florissantia quilchensis*) which are its namesake, here can be found dawn redwood, ginkgo, katsura, birch, elm, beech, sycamore, cherry, apple, hawthorn, and many other leaves. Visitors can dig for fossils and may keep three, although the Center reserves the right to retain any of scientific value.

"Terror bird" *Diatryma gigantea*, South Elk Creek, Wyoming. American Museum of Natural History, New York.

This is the first time we'll mention John Day Fossil Beds National Monument near Kimberly, Oregon, but it won't be the last. An unbroken sequence of exposed rocks that is one of the longest continuous geological records anywhere extends for 40 million years from the mid-Eocene to the late Miocene. The Eocene unit is the Clarno Formation, which is ripe with fossils of fruits, nuts, leaves (including bananas), and the bones of rhinos, brontotheres, and crocodiles. Fossils are on display at the visitor's center, and trails lead into the badlands where you may spot bones or teeth weathering out of the rock.

Fossil Butte National Monument outside of Kemmerer, Wyoming showcases fossils of leaves and fish from the lakes of the world-renowned Green River Formation. Interpretive trails wind through the badlands with trailside fossil displays. If you want to dig for Green River fossils that you can keep, there are several private quarries nearby—including Ulrich's and Warfield—where you can dig for a day for a modest fee.

The Eocene • 191

Most people don't go to Yellowstone National Park in Wyoming's northwest corner because they're looking for fossils, but there are the stumps of *in situ* petrified trees on Specimen Ridge in the northern part of the park. The trees are species of *Sequoia* and are found in the Sepulcher Formation. These trees were preserved where they stood by volcanic ash falls.

Crocodile
Crocodilus sp.,
Bridger Formation,
Wyoming. Denver Museum of
Nature & Science, Denver,
Colorado.

Similar petrified *Sequoia* trees— some 2 m (6 ft) in diameter—can be found at Florissant Fossil Beds National Monument in the mountains west of Colorado Springs, Colorado. Late in the nineteenth century, before the area was protected, many petrified logs were removed and shipped back east for display. One entrepreneur even tried to saw off one of the standing trunks, and the failed attempt is evidenced by the rusted sawblade which is still stuck in the stone tree. An easy, mostly level interpretive trail winds through the fossil forest. You won't see the leaves and insects for which the Florissant Formation is famous along this trail—they're too delicate to withstand the weather, and need to be collected from freshly-exposed rock—but the visitor's center has many fine specimens on display. Again, a few miles north of the monument there is a private quarry, Claire Ranch, where you can dig for and keep your own fossils from the same rock unit exposed in the park. When my oldest daughter was five, she found both part and counterpart of a beautiful, complete hickory leaf there.

The Eocene had seen North America cloaked in verdant forests and warm from southern tip to northern shore. But the times, they were a-changin', and the inexorable march towards an icehouse world had begun. The story of the remainder of the Tertiary is one of slowly declining temperatures and a concomitant drying of the continental interior, which would bring significant changes to North America's flora and fauna with the onset of the subsequent period, the Oligocene.

Chapter Fourteen
The Oligocene

The sun sets over an Oligocene river in Oregon. A group of oreodonts surveys the valley from an oak-cloaked bluff while a herd of rhinos comes down to the water's edge to drink. Four horses and two archaic camels have already crossed the meandering stream.

Oligocene North America

Oligocene North America was still mostly forested, but the slow descent toward an "icehouse" world had begun. We've seen that the tropical forests of the Eocene had begun to retreat toward lower latitudes, and deciduous forests largely took their place. The fauna of our continent had to adapt to these changing conditions, and this is the story which is told in Oligocene rocks.

Definition and Nomenclature

The Oligocene was not included in Lyell's original series of Epochs of the Tertiary; it was proposed in 1854 by H. E. von Beyrich to comprise strata in Germany and Belgium previously assigned to the Upper Eocene or Lower Miocene. The name means "slightly recent" and is a reference to its temporal position along the Tertiary timeline. The base of the Oligocene, or Eocene-Oligocene boundary, was set in 1992 at the extinction of the planktonic foraminifer *Hantkenina* at the base of the marl bed 19 m above the base of the Massignano quarry, Ancona, Italy, with a date of 34 mya. The Oligocene-Miocene boundary, or end of the Oligocene, was defined in 1996 as the lowest occurrence of the planktonic foraminifer *Paragloborotalia kugleri*, and near the extinction of the calcareous nannofossil *Reticulofenestra bisecta*, at the base of the magnetic polarity chronozone C6Cn.2n, 35 m from the top of the Lemmee-Carrosio section, Carrosio village, north of Genoa, Italy (whew!), which is currently dated at 23 mya. This gives the Oligocene an approximate timespan of 11 million years.

Paleogeography and Paleoclimate

North America's position on the globe has changed very little since the onset of the Oligocene: latitudinally, we were by then much where we are now; longitudinally, there has been modest westward drift concomitant with the continued opening of the Atlantic Ocean. The main effects of this have been the progressive development of the Cordillera (western mountains) and the shifting of the holarctic land connection (for faunal immigrations) from a Greenland-European route to an Alaskan-Siberian (Beringian) one.

Scallop with sponge borings, *Pecten jeffersonius*, North Carolina. Museum of Western Colorado, Fruita.

Finer-scale geographic tuning has taken place, nonetheless. By Oligocene times, the Rocky Mountains—which were originally not very high, as we've noted—had been worn down to a rolling surface called a peneplain sporting only modest relief, on the order of 1 km (3000 ft) or so. This was due to a combination of erosion and the infilling with sediment of the intermontane basins. What are now California and the Great Basin states were mostly forested and of low relief. As the North American plate slowly overrode the junction of the Farallon and Pacific plates, the incipient San Andreas fault system was born. This northward-moving fault brought many small terranes up from Mexico to meld with the California coast, and the first rumblings of volcanism in the Pacific northwest could be heard.

We've seen how, in the mid-Eocene, the separation of Australia and South America from Antarctica allowed the circumpolar current to begin to be isolated from warmer ocean waters. These plate motions continued throughout the Oligocene, with the result that the marine passages between the three continents became wider and deeper. Instead of having to rise to crest the continental shelves, both the Southern Ocean and the Drake Passage now hosted abyssal seafloors that could accommodate a deepwater current. The inceptive circumpolar stream was strengthened, becoming truly frigid. Deep-sea drill cores have brought up ice-rafted debris from Oligocene sediments in the region, attesting to the development of glaciation on Antarctica by this time.

Sabertooth *Hoplophoneus* attacking horse *Mesohippus*, White River Formation, South Dakota. North American Museum of Ancient Life, Lehi, Utah.

As waters became locked up in Antarctic ice, sea levels fluctuated and fell. The Mississippi Embayment retreated toward the Gulf of Mexico, and the continental shelves along the Eastern Seaboard were exposed. During a lowstand during the mid-Oligocene, rivers incised deep valleys in the continental shelf off New Jersey and Virginia, creating submarine canyons that have been discovered by seismic reflection. The climate of the continental interior became more seasonal as maritime

influence was reduced. Temperatures across North America declined some 8-10°C (14-18°F) from their Eocene values, perhaps in as little as a half a million years. And along with the cooling came a decrease in rainfall. These changes are reflected in the development of paleosols (ancient soils) which begin to show caliche horizons, indicating seasonal dryness for the first time during the Tertiary.

Life in the Oligocene

As one would expect, climatic changes of this magnitude had major consequences for the flora and fauna of our favorite continent. The subtropical evergreen broadleaf forests, which had grown up to 60°N during the Eocene, retreated to less than 35°N (about the latitude of Albuquerque, New Mexico and Columbia, South Carolina). Abundant palm trees along the Gulf Coast of Texas and Louisiana indicate an absence of severe frost, but colder winters began to plague the interior of Canada and the northern Great Plains. Crocodiles disappeared from Wyoming. Deciduous broadleaf forests spread, as did conifers at the higher latitudes and altitudes. The Oligocene world would still have seemed warm to us today, with forests in Alaska resembling those of the Pacific Northwest coast, but conditions were much less equable than they had been at any time since the late Paleozoic.

Pine cone *Pinus crossii*, Creede Formation, Creede, Colorado. Private collection.

With the drier conditions came new opportunities for grasses and composites (herbs with daisy-like flowers, such as sunflowers, asters, and black-eyed susans). These plants are well-adapted for seasonal drought, and fared well as open woodland mosaics began to replace closed-canopy forests across much of the midcontinent. Forest-dwelling animals became more restricted in their distribution, and forms adapted for low browsing and more open habitats appeared.

Few novel animal groups appeared in North America during the Oligocene, but there was no dearth of evolution amongst existing taxa. The oreodonts were the most diverse artiodactyls (even-toed ungulates) here at that time, followed by a modest array of entelodonts ("terminator pigs"), peccaries, and odd beasts called protoceratids that looked like small-antlered moose with nasal horns. The first radiation of archaic camelids took place, and this group continued to be moderately successful. Among the perissodactyls (odd-toed ungulates), rhinos and their relatives were the most diverse, including the hyracodont "running rhinos" and the semiaquatic amynodonts. After the extinction of the brontotheres around the Eocene-Oligocene boundary, rhinos became the largest land mammals in North America. Many of these groups went extinct by the end of the Oligocene, and tapirs retreated to Central America with their tropical forest habitat. Chalicotheres were still around, but not terribly common. Horses began to show some adaptations (longer legs, higher-crowned teeth) that foreshadow the development of savannah

Freshwater ostracods (small crustaceans), White River Formation, Pawnee Buttes, Colorado. Author's collection.

habitats during the Miocene. There was also a general trend amongst herbivorous mammals towards increasing body size, which reflects the need to process large volumes of relatively low-quality forage. Another development was an increase in the number of gregarious (herding) mammals.

These animals—particularly the large herds of oreodonts—were preyed on by the hyaenodontid creodonts and catlike nimravids, the amphicyonid bear-dogs, and a variety of early canids (true dogs). With the increase in open habitats, the canids began to radiate into the role of pursuit predators, perhaps using pack-hunting techniques against large prey, as they do today. The creodonts and bear-dogs, with their ambush-hunting style, were more suited to forests and forest edges, and this may partially account for their decline. The first mustelids (weasels and their ilk) immigrated from Eurasia and filled the small predator roles. They hunted rodents, the last few multituberculates (which were extinct by the middle Oligocene), and the newly-evolved passeriforms (songbirds). Woodpeckers, swifts, and modern ducks and swans also appeared at this time, and burrowing rodents became common, with the need to find shelter in open country. The success of many snakes follows on the heels of this increase in small field-dwelling mammals. Those voles and gophers had much to fear—the first diurnal (daytime-hunting) raptors (hawks, eagles, falcons) also appeared during the Oligocene.

Ocean life adapted by migrating equatorward as well, and coral reefs became restricted to warm, low-latitude waters. Marine plankton, especially sensitive to cold, retreated south and many fish and invertebrates followed. The major Oligocene development in sea life was the evolution of the two modern cetacean lineages—the baleen and the toothed whales—from archaeocete ancestors; whales also lost their hind limbs around this time.

Where Oligocene Fossils Are Found

Tortoise *Stylemys nebrascensis,* White River Formation, South Dakota. Private collection.

Because there was mostly erosion east of the Mississippi River, the Oligocene fossil record of eastern North America is poor, but the Brandon Lignite of Vermont reveals a plant megafossil assemblage indicative of a forest similar to those that grow on the Gulf Coast of northeastern Mexico today. Many fine shark's teeth come from the Chandler Bridge Formation of South Carolina. The Catahoula Formation of Mississippi, Louisiana, Texas, and Tamaulipas produces some beautiful petrified palm wood, as well as petrified wood of several hardwood species including oak, elm, maple, and honey locust. The Vicksburg Group, which underlies the Catahoula, contains shark's teeth and many invertebrate fossils. The famous Dominican amber from the

Caribbean island of Hispaniola was formed over about a 25-million-year span from the late Eocene through the mid-Miocene, making much of it Oligocene in age; many pieces contain perfectly-preserved insects, as well as small plant inclusions and the occasional tiny lizard.

Thanks to sediment deposition caused by outwash from the Rocky Mountains, the Oligocene fossil record of the Great Plains and West is much better. Colorado's Creede Formation produces one of the best-known Oligocene fossil floras anywhere, including spruce, fir, pine, mountain mahogany, hawthorn, barberry, and Oregon grape. The White River Formation of Colorado, Nebraska, and South Dakota produces a plethora of fossil mammals: oreodonts aplenty, camels, rhinos, entelodonts, creodonts, horses, and a variety of rodents. South Dakota's Brule Formation yields some spectacular skeletons of rhinos, horses, oreodonts, camels, entelodonts, rabbits, rodents, lizards, turtles, and snails. The John Day Formation of Oregon produces dogs, peccaries, oreodonts, horses, camels, rhinos, and rodents. A diverse mammalian fauna is known from the Sespe Formation of California. An interesting, and different, flora comes from that same state's Chicken Creek Formation: while leaves can occasionally be found (including alder, birch, maple, willow, and cottonwood), the majority of the fossil specimens are those of the winged seeds (samaras) of conifer trees, along with ostracods and some insects. The Katalla, Kootznahoo, and Kenai formations of Alaska yield fossil leaves indicating a cool, moist forest of mixed hardwoods and conifers.

Where Can Oligocene Fossils Be Seen in the Field?

There are several very nice parks scattered across the United States where you can view Oligocene fossils. The Mississippi Petrified Forest (a privately-operated park, but no collecting allowed) showcases many nice specimens of petrified trees that grew along the ancient Gulf of Mexico shoreline. This preserve, in the Forest Hill Formation, is located near Flora, Mississippi.

Badlands National Park, at Interior, South Dakota, is better-known for the erosion-sculpted hoodoos and colorful badlands topography than for its fossils, but exposures of the White River Formation there contain numerous oreodonts, horses, rhinos, pigs, and other animals. Hiking trails wind throughout the badlands, and the fossils are easy to spot because their white color contrasts sharply with the greys and reds of the surrounding rock.

Exposures of the Titus Canyon Formation in Death Valley National Park, California have produced horses, camels, oreodonts, and turtles. And finally, the Oligocene rocks exposed in John Day Fossil Beds National Monument at Kimberly, Oregon—those of the John Day Formation—yield many types of fossil leaves (including broadleaf and coniferous species) as well as horses, oreodonts, camels, rhinos, and smaller mammals. If you don't see any fossils *in situ* as you walk the trails, there are nice displays at the visitors' center.

The Oligocene was a time of transition: previously, North America had been cloaked in tropical forests; afterward, vast grasslands spread across the interior. The Oligocene watched as the great forests fragmented and began to retreat towards areas with more reliable moisture, and as the animals which roamed the continent adapted to these new conditions. For the first time since the Permian, there was glaciation over the South Pole, and it was only a matter of time until our continent saw the ice as well.

Chapter Fifteen
The Miocene

The Miocene saw savannahs spread across the interior of North America. In this snapshot of South Dakota, three gazelle-camels flee from a predatory piglike entelodont, as do a herd of horses and a long-necked llama. A lone protoceratid is unconcerned.

Miocene North America

avannah ecosystems established themselves in the North American midcontinent during the Miocene. In the modern world, the best analogue we have is the Serengeti Plain of East Africa: tree-lined watercourses winding through seas of grass, trampled and grazed by innumerable herds of horned ruminants, the tall vegetation eclipsing wily, pack-hunting carnivores and a lone bull rhinoceros while opportunistic vultures soar overhead. But this scene could have been set in Nebraska a dozen million years ago—and it was, though the makeup of the menagerie would seem strange to us now.

Definition and Nomenclature

The Miocene was one of Lyell's original epochs, the name meaning "less recent" in reference to its difference from the (following) Pliocene fauna. The start of the Miocene, or Oligocene-Miocene boundary, was again (get ready) ratified in 1996 at the lowest occurrence of the planktonic foraminifer *Paragloborotalia kugleri*, near the extinction of the calcareous nannofossil *Reticulofenestra bisecta*, at the base of magnetic polarity chronozone C6Cn.2n, 35 m from the top of the Lemme-Carrosio section, Carrosio village, north of Genoa, Italy. This currently bears a date of 23 million years ago. The Miocene-Pliocene boundary, or end of the Miocene, is defined in an equally tongue-twisting manner—at the top of the magnetic polarity chronozone C3r, ~100 kyr before the Thvera normal-polarity subchronozone (C3n.4n), and near the extinction of the calcareous nannofossil *Triquetrorhabdulus*

rugosus and the lowest occurrence of *Ceratolithus acutus*. The global stratotype for these markers was chosen in 2000 to be at the base of the Trubi Formation at Era-clea Minoa, Sicily, Italy, which is currently dated to 5 mya. This gives the Miocene an approximate duration of 18 million years.

Maple seed *Acer negundoides*, Sucker Creek Formation, Rockville, Oregon. Private collection.

Paleogeography and Paleoclimate

After the refrigeration of the Oligocene, temperatures moderated a bit, and the icecaps on Antarctica melted back some, raising sea levels worldwide. Our continent had essentially its modern coastlines by this time, but for the amount of continental shelf that was flooded. Parts of the Atlantic States were under water, as was a strip of the Pacific Coast, but the two most noticeable changes in North America's outline were the absence of Florida and the Yucatán: both were there, but consisted of submerged carbonate platforms (coral reefs) rather than dry land. The Gulf Coast in general was a bit further inland than it is today, but the outline of our continent would have been quite recognizable.

The languishing Rocky Mountains received a shot in the arm during the mid-Miocene, around 15 mya. Renewed uplift raised them to their present lofty heights as the entire western third of the continent was 'arched' up to a mile or so above sea level; downcutting of rivers associated with this led to the nascent tracings of such features as the Grand Canyon and Black Canyon of the Gunnison. This occurred about the same time as—and may be related to—the west coast of Mexico's overriding of the East Pacific Rise spreading center. Tremendous quantities of volcanic ash were spewed forth from the Cascade volcanoes, creating many spectacular fossil deposits in the West. Baja California split off from the Mexican mainland by the end of the epoch, and in the process, a strike-slip fault system was established that merged with the San Andreas to run all the way from the Gulf of California to the Queen Charlotte Islands off British Columbia. Deep rift valleys formed in southern California and filled with as much as 13 km (8 miles) of sediments. As the western fringe of the continent was (and still is) dragged northward at some 6 cm/year relative to the interior, numerous fault-block mountains formed in the Great Basin of Nevada, southern Idaho, and Utah, extending southward into central Mexico, and creating what is known today as the Basin and Range Province. Crustal stretching and thinning associated with this faulting have doubled the width of Nevada in the last 15 million years. In the process, the newborn Sierra and rejuvenated Rock-

Gomphothere *Gomphotherium phippsi*, Ogallala Formation, Nebraska. Denver Museum of Nature & Science, Denver, Colorado.

ies created rain shadows that dried out the interior of the continent, contributing to vegetational changes that we will discuss a bit later. The Miocene also saw North America's westward journey take it across a mantle "hot spot," causing the massive outpourings of lava that form the flood basalts of the Columbia River Plateau and Snake River Plain. This hot spot today rests under Yellowstone National Park, and is the source of the geothermal energy that powers that region's famous geysers. Where the Pacific plate dives beneath the North American plate, Alaska's Aleutian island chain was formed—and is still the site of volcanic activity.

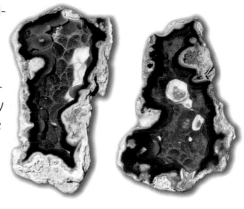

Agatized coral (sectioned), Tampa Bay, Florida. Author's collection.

One large rift deserves particular mention: the Rio Grande Rift, which begins in the mountains of central Colorado and bisects New Mexico before heading mostly southeastward along the Texas/Chihuahua-Coahuila border. This Miocene "failed rift" (so-called because the continent did not ultimately split in two along it) has determined the course of the Rio Grande River.

South of the border, the lands of Central America and the Caribbean were taking shape. Much of North America south of Mexico is merely a mixed mass of stuck-together island terranes that formed from Tertiary volcanism along the leading edge of the North American plate. As the raft which bore our continent overrode the Pacific plate, these terranes got mashed together, forming the rugged lands we see today. Miocene and later volcanism and reef building created the sprinkling of Caribbean islands.

After the modest warming of the Early Miocene, temperatures began to decline again. This was, again, related to the extent of glaciation on Antarctica; around 16 mya, the East Antarctic ice cap began to form. The exact impetus behind the advance of the glaciers is unclear, but in addition to continued development of the Antarctic Circumpolar Current, the Norwegian Basin between Greenland and Scandinavia was growing wider as separation continued along the Mid-Atlantic Ridge. This may have allowed the North Atlantic Deep Water to flow south from the Arctic Ocean for the first time, pushing the boundary between tropical and cold waters to higher southern latitudes. The closer proximity between warm, moist waters and the south polar landmass may have resulted in increased snowfall and thus growth of the ice sheets.

Gazelle-camel *Stenomylus hitchcocki*, southeast of Agate, Sioux Co., Nebraska. American Museum of Natural History, New York.

This later Miocene cooling trend, coupled with the rain shadow effect of the rising Sierra, Cascades, and Rockies, brought less and less rain to the interior of North America. Decreases in absolute rainfall as well as seasonal drought plagued the Great Plains and Prairie Provinces; the southwestern U.S. and much of Mexico became truly arid for the first time since the Jurassic. As we shall see, this had great consequences for the flora and fauna of our continent.

Life in the Miocene

As aridity spread across the southwestern reaches of North America, many areas of Mexico and the Great Basin developed juniper-sagebrush ("high desert"), sclerophyll, or chaparral vegetation. Coniferous forests dominated at high elevations in the mountains, as they do today. In the midcontinent, the trend throughout the Miocene is to increasing presence of grasses and annual herbs (especially composites—daisy family) in the flora. These plants are well-adapted to seasonally dry conditions, and to the greater frequency of wildfires that accompany them. This was, however, a gradual change; Early Miocene ecosystems in the Great Plains indicate savannah vegetation—mainly grasslands but with up to 40% tree cover in isolated groves and along watercourses ("gallery forests"). Slowly, the trees became confined to the banks of frequently-ephemeral streams, and by the Late Miocene, true prairies or steppes—seas of grass punctuated only rarely by isolated trees—occupied the interior of North America. Grasses survive drought by going dormant, and aerial dispersal of their tiny seeds is a big advantage on open, windy plains. Many species spread by ground-level or underground runners and the blade of the grass grows throughout its length, allowing them to suffer intense grazing pressure from herbivores and still survive.

Elsewhere on the continent, remnants of vegetation from warmer, moister times made their last stand. Miocene plant-bearing localities are rare in eastern North America, but the little evidence that we have (mostly pollen) indicates that these regions were covered with a broadleaf deciduous forest, generally similar to what is presently found in Georgia or Florida but also with a few relict species now confined to Mexico and Central America. The Gulf Coast hosted extensive swampy regions and bayous. On Devon Island in the Canadian Arctic, fossils indicate a mixed conifer/northern hardwood forest such as grows south of the boreal forests today (e.g., in Michigan's U.P. or around Montréal). The North Slope of Alaska bore coniferous forests right up to the shores of the Arctic Ocean, in areas that now support only tundra. Moist forests on the West Coast were confined to seaward of the Sierra and Cascades.

Shark teeth
Carcharodon megalodon,
river deposits,
Florida.
Private collection.

While terrestrial forests were retreating, a new type of "forest" habitat appeared during the Miocene: the kelp forest. These dense groves of marine macroalgae are common in temperate waters off the Pacific Coast. They provide shelter for a multitude of fishes and invertebrates, and feeding grounds for such animals as sea otters (which are first known from this timeframe) and sea cows. Seals, sea lions, and walruses all also appear during the Miocene epoch, and huge relatives of the Great White Shark some 15 m (50 ft) long—*Carcharodon megalodon*—plied the waters, occurring in great numbers proximal to the whale calving grounds of Chesapeake Bay. Many bones of immature whales are found there bearing scratch marks from the teeth of this voracious shark.

Reef faunas differed little from those of the preceding Oligocene or the succeeding epochs of the Cenozoic, being essentially modern in their aspect. The major change in the reef environment during the last 20 million years has not been in its faunal makeup, but in its extent: as the globe and oceans cooled, coral reefs became more and more restricted in their distribution, retreating towards equatorial latitudes. For North America, this means that they are only found off Florida, in the southern Gulf of Mexico, throughout the Caribbean, and in the Pacific from Mexico to Panama.

As we've noted, by the mid-Tertiary the primary holarctic connection shifted from an eastern route via Greenland to an Alaskan-Siberian one. The Bering Strait today is only 55 miles (88 km) wide, and at sea level lowstands, North America and Asia are connected by a broad land bridge called Beringia. As widening of the Atlantic caused the Norwegian Sea to become a barrier to floral and faunal migrations, the Beringian route opened up, such that many animals could still roam freely betwixt the northern continents. The Miocene, therefore, saw wave after wave of immigrants enter our continent from Eurasia.

Two groups of proboscideans (pachyderms) entered North America at this time: the four-tusked gomphotheres and the two-tusked mastodons. Mastodons were forest browsers and were particularly common across Canada and in the forested Great Lakes region. The gomphotheres ranged widely across the continent and seem to have preferred savannah riparian habitats, where they could use their lower tusks to dig up aquatic plants.

Cervids (deer) appear in North America during the Miocene, and are also immigrants from Eurasia. Bovids (sheep, goats) arrived as well. Rhinos were still common

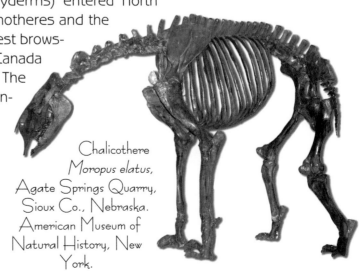

Chalicothere
Moropus elatus,
Agate Springs Quarry,
Sioux Co., Nebraska.
American Museum of
Natural History, New
York.

Pocket mouse *Perognathus* sp., Cherry Co., Nebraska. American Museum of Natural History, New York.

across the continent, but with the extinction of the hyracodonts ("running rhinos") and semiaquatic amynodonts at the end of the Oligocene, only rhinoceratid (true) rhinos were left. Camelids and peccaries remained numerous; one camel, *Aepycamelus*, developed a very long neck and may have been the ecological equivalent of the giraffe. Oreodonts were on the wane (possibly due to loss of their woodland habitat), and didn't survive the end of the Miocene. The few remaining protoceratids ("horned moose") retreated to the Gulf and Caribbean coast. Chalicotheres hung on but went extinct at the close of the epoch. Antilocaprids (pronghorns) are a native North American group which enjoyed tremendous success during the Miocene, evolving a number of species with variously-shaped horns. They are the only ungulates to have permanent bony horn cores covered by a deciduous keratin sheath; they are also the speediest mammals on our continent, reaching some 60 mph (100 kph). As arid habitats spread across the West, the dry-adapted pronghorns expanded their range as well, reaching a distribution that spanned from central Canada to northern Mexico. Horses diversified greatly, with as many as 12 different species coexisting at one time in the mid-Miocene. As the epoch progressed, however, diversity declined again, with browsing (leaf-eating) species suffering the heaviest losses while grazing (grass-eating) ones flourished.

Extinct Great White Shark *Carcharodon megalodon* (original teeth, reconstructed jaws), Beaufort, South Carolina. Private collection.

Felids (true cats) arrived from Eurasia and replaced the few remaining nimravids as ambush-style "hypercarnivores." The amphicyonid bear-dogs declined in abundance—going extinct by the end of the epoch—as ursids (true bears) became established and canids (dogs, wolves, coyotes) diversified, including some bone-cracking, hyena-like forms. Entelodonts ("terminator pigs") made their last stand early in the epoch. They were very small-brained compared to the abovementioned predators, and this may have been their downfall.

Sloths colonized our continent during the mid-Miocene, but their appearance is somewhat of a mystery. They clearly immigrated from South America, but how they managed it, without a land bridge at Panama, is problematic; perhaps they arrived on mats of floating vegetation. (This isn't as bizarre as it sounds: in tropical regions today, hurricanes can generate rafts of tangled vines and branches large enough to harbor monkeys and other substantial-sized animals.) The sloths populated the warmer regions of our continent, spreading northward through Mexico as far as Florida and the southern Great Plains. Raccoons invaded North America at this time, and there was a wide radiation of mustelids (weasels, skunks).

Throughout the Tertiary, three main trends can be recognized time and again in diverse lineages of mammals: 1) increasing brain size; 2) specialization of the teeth for a particular diet; and 3) specialization of the limbs for the appropriate habitat. A good example of this may be seen in the ungulates. Both artiodactyls (even-toes) and perissodactyls (odd-toes) show the trends of increasing adaptations to open habitats as the Tertiary progressed. The spread of savannah and prairie from the Miocene onward led to a shift from browsing to grazing diets among many ungulates. Grasses contain small pieces of silica, called phytoliths, which make them very abrasive; an animal nipping close to the ground is also more likely to ingest dirt and grit than one feeding on the leaves of trees or shrubs. The consumption of abrasive foodstuffs wears down the teeth very rapidly, so there is a trend for later ungulates to develop higher-crowned teeth than their ancestors, with more complexly-folded enamel patterns. The tooth continues to grow until the animal is middle-aged, keeping up with wear, so that it will last a lifetime. Open habitats also allow large animals to forage over wider territories, so longer legs become an advantage for efficient travel. Savannahs and prairies offer precious few places to hide from predators, so those long legs are also needed to outrun their pursuers. The classic "eohippus-to-*Equus*" sequence of horses on display in many museums shows these trends particularly well (although the full family tree of horse evolution is 'bushier' than this one lineage suggests). Ironically, this progression was first noticed by Sir Richard Owen, a staunch creationist. It was later used by Thomas Henry Huxley to argue for Darwin's theory of evolution.

Additionally, the increasing dominance of grassland habitats led to three other trends within the ungulates: progressively increasing body sizes, the development of skull ornamentation (horns, antlers), and herding behavior (evidenced by monospecific bonebeds in the fossil record). All of these correlate with a life in open environments. Grasses are low in nitrogen, thus inherently less nutritious, than most other leaves; a grazing animal must therefore consume copious quantities of fodder, encouraging the increase in body size necessary to accommodate a large gut. Large size in and of itself is also a powerful deterrent to predators (no carnivore today has a prayer of

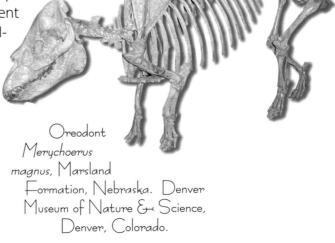

Oreodont *Merychoerus magnus*, Marsland Formation, Nebraska. Denver Museum of Nature & Science, Denver, Colorado.

bringing down a healthy, full-grown elephant). Horns and antlers are not only useful for sexual display and species recognition, but they are a liability in forests and not so out in the open. Herding behavior is a manifestation of the "there's safety in numbers" philosophy when attempting to avoid predators in a habitat where there's no place to hide.

As herbivores developed long legs for fast running and efficient travel, there is a trend amongst the carnivores to become more efficient pursuers as well. A time lag between the development of cursoriality in herbivores and in carnivores (cats, dogs) suggests that the need for long-distance foraging was the initial impetus behind the trend, and that the ability to outrun speedy predators was developed secondarily. Cats—including the formidable sabertooths, which were the dominant feline carnivores in North America from the Miocene through the Pleistocene—tend in general to be ambush predators, but can put on an astonishing burst of speed over a short distance. They first appear in the North American fossil record during the Late Miocene, immigrants from Eurasia. Canids (dogs, wolves) tend to be endurance predators, hunting in packs and chasing their prey until it becomes exhausted. An herbivore that can run fast and long has a big advantage in both these situations.

Sycamore leaf *Platanus dissecta*, Sucker Creek Formation, Sheaville, Oregon. Private collection.

Mammals were not the only group whose evolution was affected by the increasing dominance of grassland ecosystems. The success of snakes closely parallels an increase in the numbers and diversity of small rodents that live in the grasslands (imagine a prairie dog town, or think of how many mice live in an open field). Diurnal raptorial birds such as hawks and eagles largely exploit this same resource. The Miocene also saw the evolution of humongous vultures called teratorns, which could range widely over the plains in search of carrion.

Where Miocene Fossils Are Found

The few Miocene formations exposed in the eastern states are mostly marine, and confined to nearshore areas that were flooded during sea level highstands. The Calvert, Choptank, and St. Mary's formations of Maryland and Virginia are justly famous for their beautiful snails, clams, and sand dollars (a type of urchin), as well as for the plethora of different shark's teeth found there. The Calvert Formation represents the most extensive Miocene fauna known from the East Coast. Miocene marine formations are also exposed in the Gulf Coast states. A fauna containing mastodons, rhinos, llamas, horses, pronghorns, fish, turtles, and alligators has been reported from Louisiana's Pascagoula Formation.

We have noted how ash fallout from the Cascade volcanoes repeatedly blanketed the western states and provinces, creating an array of stunning fossil deposits, some of which preserve even delicate bird bones in perfect three-dimensionality. A tremendous amount and variety of petrified wood comes from various volcanic ash layers sandwiched between the Columbia River basalt flows in Washington State, and there is even the unusual hollow fossil of a rhinoceros embedded within one of the lava flows, called the "Blue Lake rhino." This singular fossil was created when an unfortunate rhinoceros was immolated as he stood in the path of the molten basalt. The rock hardened around the rhino, leaving a rhino-shaped cavity that even contained a few bones when it was first discovered early in the 20th century.

In the Nevada desert, exposures of the Middlegate Formation produce spectacular plant fossils, most of them leaves and acorns of the live oak but also including giant *Sequoia*, willow, fir, maple, and spruce. The overlying Monarch Mill Formation yields a rich vertebrate fauna comprised of oreodonts, camels, rhinos, pronghorns, weasels, rabbits, squirrels, and beavers. An abundance of leaves can be found in the same state's Buffalo Canyon Formation, including those of live oak, birch, cypress, juniper, cottonwood, willow, alder, hornbeam, hickory, and black walnut. California's Mehrten Formation is another prolific producer of fossil leaves. Bears, deer, horses, camels, rhinos, bear-dogs, cats, and weasels are all found in Oregon's Mascall Formation. Western Nebraska's Harrison Formation yields small camels, rhinos, entelodonts, chalicotheres, and oreodonts; the slightly younger Batesland Formation of South Dakota produces bones of rhinos, camels, birds, entelodonts, chalicotheres, and rodents. Gomphotheres and rhinos come from the Ogalalla Formation of Colorado and Nebraska.

California's Monterey Formation, widely exposed in coastal areas of the state, preserves the remains of one of the earliest-known kelp forests. In addition to fossils of these macro-seaweeds, spectacular crabs and the bones of a variety of ancient whales and dolphins are found there. Whale bones and shark's teeth have also been reported from the Topanga Formation of the same state, and the Round Mountain Silt Formation is well-known for shark's teeth as well as the fossils of bony fish and turtles. The richly fossiliferous Astoria Formation of Oregon, also a marine unit, is known to produce clams, snails, and scaphopods (another type of mollusc). South of the border, the Tuxpam Formation of Veracruz, Mexico is well-known for a plethora of bivalves, gastropods, and echinoderms, as well as a few crustacean fossils.

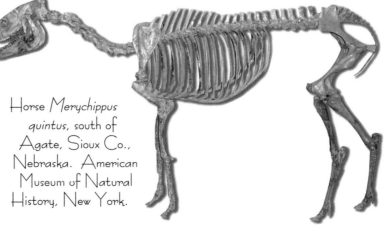

Horse *Merychippus quintus*, south of Agate, Sioux Co., Nebraska. American Museum of Natural History, New York.

Where Can Miocene Fossils Be Seen in the Field?

Miocene fossils are the beneficiaries of many public preserves from coast to coast. Probably the most famous site in the east is Calvert Cliffs State Park at Marbury, Maryland, on the shores of Chesapeake Bay. Here rocks of the Chesapeake Group (Calvert, Choptank, and St. Mary's formations) are exposed along the public beaches. Although this park gets a lot of visitors and can be pretty picked over at times, storms expose new fossils all the time, and the rules allow you to collect and keep any shark's teeth and clams, snails, or other invertebrate fossils you find.

A recent (May of 2000) serendipitous find was made by a road construction crew near Gray, Tennessee, and has been set aside as the Gray Fossil Site. The sediments consist of an unnamed unit of Miocene fill in a sinkhole surrounded by the Ordovician Knox Group. The sinkhole harbored a pond that preserved the remains of residents as well as those of visiting animals that came to drink and may have drowned. Bones of rhinos, tapirs, gomphotheres, camels, peccaries, weasels, badgers, turtles, frogs, alligators, and fish have all been recovered. A new museum building opened in late summer 2007 to showcase fossils found at the site.

Ashfall Fossil Beds State Historical Park near Royal in northeastern Nebraska preserves the remains of scores of rhinos that perished in a volcanic ash fall. The fossils are found in an ash bed in the Ash Hollow Formation. The animals apparently perished from inhalation of the choking ash, and were buried and preserved in three dimensions and usually fully articulated. In addition to the rhinos, the bones of horses, dogs, camels, deer, birds, and turtles have been found; some of these specimens are among the best-preserved of their kind known anywhere. Some of the female rhinos even enclose unborn young in their body cavities! Inside the "Rhino Barn" you can watch as paleontologists continue to excavate the fossils, if you visit during the summer months.

Pronghorn relative *Ramoceros osborni*, Cedar Creek, Colorado. American Museum of Natural History, New York.

Agate Fossil Beds National Monument in western Nebraska, near Harrison, is also well worth a visit. Here the Harrison Formation yields numerous rhinos, chalicotheres, entelodonts, and other Miocene animals. One of the most interesting and unusual (I thought) is *Daemonelix*, or the "Devil's corkscrew." These are the petrified, screw-shaped burrows of a terrestrial species of beaver—some six or more feet deep. It wasn't known for a long time what animal excavated the burrows, until the skeleton of the beaver was found in the bottom of one. The spacing of its incisor teeth perfectly matched the chisel marks left from excavation of the burrow. There is a nice example of a *Daemonelix* on display in the museum, and more of them can be seen *in situ* if you hike the one-mile (1.6 km) *Daemonelix* Trail. Apparently these beavers lived colonially much as prairie dogs do today, and the trail takes you to one of their "villages."

Ginkgo Petrified Forest State Park, near Vantage, Washington, preserves the remnants of a fossil forest that grew atop one of the Columbia River basalt flows. The wood occurs in the Vantage Interbed between basalt layers. *Ginkgo*, elm, *Sequoia*, Douglas fir, oak, maple, and other trees grew along the shores of ancient Lake Vantage, into which they fell after death and became waterlogged. This waterlogging prevented them from getting incinerated when the next lava flow came and sealed the area over. Although many nice stump specimens can be seen along the walking trail, sadly, they must be covered with locked gratings to prevent vandalism, which ruins one's photos.

If you're in this part of Washington, you might want to try checking out the Blue Lake rhino, although it's in a rather obscure location and not the centerpiece of any park. Drive northeast from Vantage, past Ephrata and Soap Lake, to Dry Falls. From there ask directions to Blue Lake; when you get there, look for Laurent's Sun Village Resort. This is "rhino city" with newspaper clippings and rhino reports tacked up all over the walls. They'll rent you a boat to get you to the base of the scree slope above which lies the cavernous rhino. The fossil is marked with a white spot on the cliff face above the reservoir and access is by a steep trail, but when you get there you're rewarded with

Beaver burrow *Daemonelix*, Harrison Formation, Agate Fossil Beds National Monument, Nebraska.

one of the most unusual fossils on the planet—and one you can actually crawl inside of, if you're inclined. Don't expect any bones, though; they've been curated at the University of California, Berkeley, since the 1940s.

Camel *Procamelus grandis*, Cherry Co., Nebraska. American Museum of Natural History, New York.

We've mentioned John Day Fossil Beds National Monument before. Two Miocene-aged units are exposed in the park: the Mascall and Rattlesnake formations. The Mascall is the older of the two units (15-12 mya) and contains fossils of camels, horses, deer, bears, weasels, gomphotheres, bear-dogs, cats, and true dogs. Fossils in the overlying Rattlesnake Formation (8-6 mya) aren't generally as well-preserved as in the Mascall, but include horses, sloths, rhinos, pronghorns, camels, peccaries, bears, and dogs. If you find yourself in the Kimberley, Oregon area, be sure to stop by.

Petrified wood (bald cypress), Washington. Private collection.

Anza-Borrego Desert State Park in southern California protects two distinct units containing Miocene fossils. The Deguynos Formation has produced the bones of walruses and dolphins as well as shark's teeth and oysters, whereas the Latrania Formation yields mostly shark's teeth and molluscs. If you go, be properly prepared for some desert hiking, and ask about trail conditions at the visitor's center.

The Miocene saw savannahs and steppes spread out across the interior of our continent in response to climatic cooling and drying. The great broadleaf forests were now split in two by an ocean of grass. A world away, Antarctica was shrouded in ice—a change that had worldwide repercussions. The vast herds of large animals that roamed across North America would have to adapt to deteriorating conditions, or perish.

Chapter Sixteen
The Pliocene

Cooler, drier times presaged the Ice Ages in Pliocene North America. Arid conditions came to Chihuahua where a glyptodont tries to drink from a mud-caked arroyo, as do a pair of mastodons. A horse nips at sparse, low-growing greens, but three pronghorns are in their element. Overhead, a large vulture waits for animals to succumb to the drought.

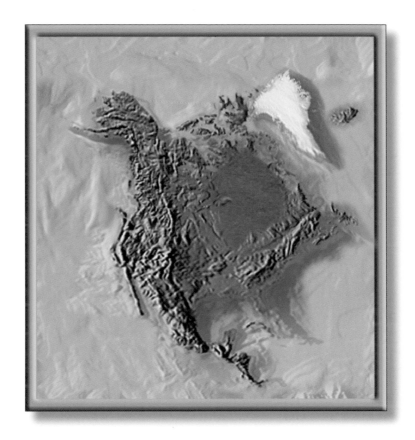

Pliocene North America

We've seen global climates deteriorate during the latter half of the Miocene, and this trend would continue throughout the Pliocene. Antarctica was permanently ice-bound by now, bringing a chill to weather the world over. The three million years of the Pliocene would see icecaps established in northern regions as well, as plants and animals scrambled to adapt to the changing conditions or retreat towards the equator.

Definition and Nomenclature

The Pliocene is the last epoch of the Tertiary Period. It is the third of Lyell's original Tertiary epochs. The name means "more recent," in comparison to the more archaic faunas of earlier in the period. Originally, Lyell's Pliocene encompassed all of what we would now consider Pliocene and Pleistocene time, under the guise of "older Pliocene" and "younger Pliocene"—which Lyell himself revised to reflect the current nomenclatural scheme in 1839. The beginning of the Pliocene, or Miocene-Pliocene boundary, was defined in 2000 at the top of the magnetic polarity chronozone C3r, ~100 kyr before the Thvera normal-polarity subchronozone (C3n.4n), and near the extinction level of the calcareous nannofossil *Triquetrorhab-dulus rugosus* and the lowest occurrence of *Ceratolithus acutus*. This has been dated to 5 mya with a global stratotype designated at the base of the Trubi Formation, Eraclea Minoa, Sicily, Italy. The Plio-Pleistocene boundary (close of the Pliocene, and of the Tertiary) was set in 1985 just above the top of the magnetic polar-

Sand dollars *Dendraster gibbsii*, Etchegoin Formation, Kings Co., California. Author's collection.

ity chronozone C2n (Olduvai) and the extinction level of the calcareous nannofossil *Discoaster brouweri* at the top of sapropel layer 'e,' Vrica section, Calabria, Italy, with an age of 1.8 mya, fairly precisely derived from astronomical cycles in the sediments. Just above this horizon are the lowest occurrence of the calcareous nannofossil *Gephyrocapsa* spp. and the extinction level of the planktonic foraminifer *Globigerinoides extremus*. The Pliocene, therefore, spans just a little over 3 million years, making it the shortest epoch of the Tertiary.

Paleogeography and Paleoclimate

Given its short timespan, relatively little change took place in the geography of North America during the Pliocene. Trends that had been established during the Miocene continued: the rifting of the Basin and Range Province, the rise of the Sierra, the seaward retreat of the coastlines. The West Coast mountain ranges rotated slowly clockwise—as if on a door hinged in Puget Sound—to attain their present north-south orientation. Volcanism in the Cascades was on the wane, possibly due to the nearly-complete subduction of the Farallon Plate; Mt. St. Helens is one of only a handful of remaining active volcanoes in the Pacific Northwest. Our continent continued to drift slowly westward, leading to the cessation of basalt emplacement on the Snake River Plain and the settling in of the magma hotspot under the Yellowstone region. An explosive Pliocene eruption blasted out a huge caldera there, forming what is today Yellowstone Lake.

Far to the south, magma upwelling beneath the Four Corners area uplifted the Colorado Plateau more or less as a single block, adding about a mile (1.6 km) to its elevation. Rivers such as the Colorado and the San Juan, which had sculpted meandering valleys across the broadly level Miocene land surface, downcut voraciously as the uplift occurred, preserving their former courses. This is why deeply incised meanders—rather than the path of least resistance—characterize the topography of the Grand Canyon and the Goosenecks of the San Juan. Miocene drainage patterns are superimposed on much of the Rocky Mountain region, as uplift and erosion competed to stay in balance with one another.

Proboscidean jaw *Stegomastodon mirificus*, found in load of gravel delivered to Albuquerque home.

New Mexico Museum of Natural History, Albuquerque.

Remnants of forests clung to the coasts of Antarctica and southern Greenland, but the ice extended its clutching grasp. The Antarctic ice cap continued to grow until it enveloped the whole continent, refrigerating the entire globe. During the mid-Pliocene (around 3.5 mya), South America nudged up close to Panama and a land bridge was formed between our continent and our southern neighbors. This closely corresponds in time with the initiation of glaciation in the northern hemisphere; although the exact impetus is obscure, it may be related to shifting ocean currents as the Panamanian Land Bridge cut off the Caribbean from the Pacific and disrupted circum-equatorial ocean circulation. As icecaps began to form in Greenland and northern Canada, we entered our present climatic regime of orbitally-controlled Ice Ages. Cold air can hold far less moisture than warm air, so more aridity accompanied the chilling trend.

Life in the Pliocene

As the climate progressively cooled and became drier, trees in the Great Plains became fewer and farther between, causing the expansion of prairie and steppe ecosystems. The spread of grasslands was abetted by an increase in seasonal droughts and wildfires, as well as an absolute decrease in rainfall. True deserts developed in Mexico and the southwestern United States, and the last remnants of tropical forests retreated to the Caribbean shores. Temperate woodlands remained in the eastern third of the continent, with mainly coniferous forests in the western mountains. The great forests of the north, composed chiefly of conifers and hardy broadleaf species like alder and birch, still carpeted the shores of the Arctic Ocean early on, but were forced to retreat as the ice caps crawled slowly across their lands. The first northern tundra vegetation developed between 2 and 3 million years ago in areas proximal to the ice sheets. It is unclear precisely when we entered the oscillating glacial-interglacial cycle, but it is thought to have been sometime prior to the close of the Pliocene. Around 3 mya, ice-rafted cobbles begin to appear in northern Pacific and Atlantic sediments, indicating some degree of continental glaciation.

Animal migrations continued both into, and out of, North America. Following the trail blazed by the sloths, other animals such as armadillos, glyptodonts (giant armadillo relatives), opossums, and huge, fast-running phorusrhacoid "terror birds" made the journey north across Panama—these last reminiscent of the diatrymas of the Eocene, but separately derived. Most of these animals preferred warm climates and did not extend their range much past the Gulf of Mexico coast, although giant ground sloths sortied north as far as the central Great Plains. Numerous animals from North America invaded the southern lands as well, resulting in massive extinctions amongst the endemic South American fauna. Beringia, too, was wide open, and any animals that were not deterred by the cold could migrate freely between Alaska and Eurasia. Canids (dogs, wolves), thus far a native North American group, colonized the Old World via this route, as did horses and camels. The first moose and elk arrived from Asia along this northern corridor.

Alligator scute
Alligator mississippiensis, Pinecrest Formation, Charlotte Co., Florida. Author's collection.

Mako shark tooth
Isurus hastalis, Beaufort Co., South Carolina. Author's collection.

Amongst the stay-at-homes, the first one-toed horses appear, and by the end of the Pliocene, only the modern genus *Equus* was left. Rhinos vanished from North America early in the epoch, probably due to loss of their savannah habitat. Deer became extremely common as most of the archaic artiodactyl groups, including the last protoceratids, died out. Camels and pronghorns suffered huge losses in diversity; the ones that did survive were generally the most hypsodont members of the group (those with the highest-crowned teeth, indicating that they were well-adapted for a diet with a large proportion of grasses and low-growing, gritty weeds). Gomphotheres disappeared as savannah turned to steppe, but the mastodons held on, apparently more adaptable in their food requirements.

Whilst cursoriality (fast running) had evolved amongst herbivores during the Miocene, we've noted that it took a while for predators to catch up. There was a wide radiation of modern canids, especially foxes and wolves, during the Pliocene. These predators have long limbs for speed but also possess great endurance: they chase their prey until it becomes exhausted. The hyena-like canids went extinct at this time, and the carnivore fauna took on a fairly modern look, with the notable exception that great sabertooth cats still prowled the landscape.

Where Pliocene Fossils Are Found

Beautiful snails can be collected in Florida from the Arcadia Formation. The remains of a variety of sea turtles have been reported from that same state's Bone Valley Formation, in addition to an array of sharks, rays, bony fishes, birds, walruses, seals, sea otters, sea cows, and whales. A similar fauna is known from North Carolina's Yorktown Formation, with fossils of turtles, birds, fish, and whales all being common. This unit also contains nice invertebrates including corals, clams, snails, and sea urchins. The Hawthorne Formation of South Carolina produces many nice shark's teeth. Some bear and horse bones have been reported from Louisiana's Citronelle Formation.

Pliocene deposits are scarce in the Midwest, but the Pipe Creek Sinkhole near Swayzee, Indiana deserves mention. The collapse of a limestone cave carved out of Silurian limestone during the Pliocene created a water-filled sinkhole here (similar to the Miocene-aged Gray Site in Tennessee). This deposit has yielded the bones of camels, bears, peccaries, beavers, turtles, snakes, and numerous leopard frogs. Instead of being scoured away by the Pleistocene glaciers, this sinkhole was covered over with glacial till and thus preserved as a unique "time capsule."

Hagerman horse *Equus simplicidens*, Glenns Ferry Formation, Hagerman Fossil Beds National Monument, Hagerman, Idaho.

The Rexroad Formation of Kansas preserves the delicate bones of passerine songbirds, as well as those of many mammals and reptiles. The Mio-Pliocene Ogallala Formation, exposed in western Kansas and across much of the High Plains, produces an interesting flora consisting mostly of nutlets and seeds of grasses and composites with a large number of hackberry fruits. Idaho's Glenns Ferry Formation is best-known for its fossil horses, but also contains the remains of turtles, birds, and rodents.

California has quite an extensive Pliocene fossil record. The San Diego Formation, a marine unit well-exposed in the southernmost part of the state and south into northern Baja California, contains various invertebrate fossils including those of clams, snails, scaphopods, crustaceans, and sea urchins; shark's teeth can be found as well. The bones of larger marine mammals including baleen and toothed whales, fur seals, walruses, and sea cows are also known from this unit. Inland, in the Owens Valley, the Coso Formation is known for its horses, llamas, and other mammal fossils. The Palm Spring Formation has yielded the bones of many large mammals, including gomphotheres and sloths, plus some nice petrified wood. The San Joaquin Formation produces some beautiful oysters, snails, pectens, and sea urchins, and the Tulare Formation is said to yield the largest array of Pliocene freshwater molluscs of any deposit on the West Coast.

Sand dollar, Baja California, Mexico beach. Private collection.

Where Can Pliocene Fossils Be Seen in the Field?

There are few preserves set aside to protect Pliocene fossils—not because they are unimportant, but because the short duration of the epoch makes such deposits relatively scarce. All are in the West.

Possibly the most famous Pliocene fossil park is Hagerman Fossil Beds National Monument at the tiny berg of Hagerman, in southeast Idaho. These exposures of the Glenns Ferry and Tuana formations record a semiarid, high-altitude landscape that received about twice the rainfall that it does today. A vibrant community of beavers, muskrats, otters, frogs, snakes, waterfowl, mastodons, sabertooth cats, camels, pronghorns, deer, and ground sloths lived in the valley, but the single most famous fossil (found in stunning numbers) is *Equus simplicidens*, the Hagerman Horse. Unfortunately, the fossil exposures are not easily accessible to visitors, but there are lots of nice bones on display at the visitors' center including a full-mounted skeleton of a Hagerman Horse. One thing I liked about this tiny museum is that many of

Gastropod, Port Richey, Florida. Author's collection.

The Pliocene • 221

the fossil bones on display are fragmentary, so you can get a good idea about what scientists find in the field; sometimes, looking at nothing but fully-restored animals in museums, you get a distorted idea of what fossils really look like when they're first unveiled.

Anza-Borrego Desert State Park in southern California contains outcrops of the Palm Spring Formation that contain fossils of petrified wood and also horses, gomphotheres, boars, sloths, wolves, bears, and sabertooth cats. You may remember from the last chapter that this park protects Miocene rocks as well. Be prepared for some hiking, and avoid the midsummer's midday heat.

Bottom and top view of cowrie shell, river deposit, Florida. Note the boreholes from a predatory gastropod, a likely cause of the cowrie's demise. Author's collection.

Another rather out-of-the-way protected area is the Coso Range Wilderness on Bureau of Land Management (BLM) land at the south end of the Owens Valley, California. Here, in the shadow of the majestic Sierra, can be found outcrops of the Coso Formation that contain fossils of horses, rabbits, packrats, bear-dogs, llamas, peccaries, mastodons, and bears. This is an undeveloped wilderness area, so don't expect an air-conditioned museum building nearby, and no fossils may be collected without a permit.

Finally, the privately-owned Pliocene age Petrified Forest at Calistoga, California showcases an extensive area of petrified trees. This might be your best bet besides Hagerman if you're unable or unwilling to brave rigorous desert hiking conditions.

The Tertiary closed with the end of the Pliocene, as the Earth headed into the erratic climate fluctuations of the Quaternary. Cool, dry climates became cold, arid, and windy with the onset of the Pleistocene glaciations. North America became bound in ice for the first time in a billion years as the massive glaciers crept southward from centers in Greenland and Hudson Bay, all the way to the banks of the Ohio. It was into this forbidding landscape that primates—long extinct here—returned.

The Quaternary Period

Spruces and swamps covered Michigan during an interglacial between advances of the ice. Two mastodons enjoy the autumn sunshine before an approaching storm.

By now we've journeyed through the vast majority of geologic time, and seen the world evolve—geographically, climatically, biotically—into one very close to that which we know today. Nonetheless, the last two million years still contain exciting developments in the history of our continent—not the least of which is the repeated advance and retreat of ice sheets as much as two miles (3 km) thick.

The Quaternary was named in 1829 by Jules Desnoyers based on young rocks in the Seine Basin of northern France, up to and including the present. The name reflects that it is the "fourth age," after the Tertiary Period. The Quaternary, like the Cenozoic Era of which it is a part, is comprised of two very unequal subdivisions, the first of which occupies almost the entire timespan; Quaternary time consists of the Pleistocene Epoch, which lasts for 99% of its duration, and the Holocene Epoch, which is the last eleven-and-a-half millennia (the time since last retreat of the ice sheets). This is really a rather arbitrary and anthropocentric distinction, because it gives special consideration to the Holocene merely because it is the time in which we are living, and civilization has arisen—when from a geologic standpoint, the Holocene is merely another unremarkable interglacial in between extended glacial spells. Moreover, since the Holocene occupies only 1% of Quaternary time, the terms Quaternary and Pleistocene are practically synonymous. Nevertheless, because the developments of the last 11,500 years are both extremely important and likewise well-known, we'll devote a separate chapter to each of the epochs of the Quaternary. And we'll look a little bit ahead as well, to see what lessons from the past we can apply to our future.

Ok, let's go. Can you hear the wind howling? Zip up your parka—it's pretty cold out there...

Chapter Seventeen
The Pleistocene

The ice crawled south into Nebraska during the Pleistocene like a giant amoeba centered on Hudson Bay. A herd of mammoths is undeterred and browses on lush grasses and tundra shrubbery at the foot of the glacial wall. A lone caribou male watches them from a distance.

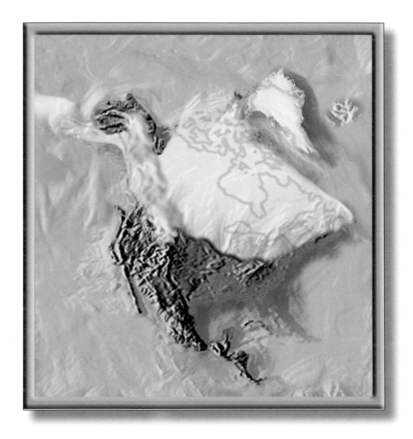

Pleistocene North America

ammoths. Mastodons. Sabertooth cats. Giant ground sloths. And ice sheets more than a mile thick. These things immediately come to mind when picturing the North America of the past two million years. The 99% of Quaternary time encompassed by the Pleistocene Epoch is not inaccurately also called the "Ice Age," and megafauna such as these were some of our continent's most obvious denizens at that time.

Definition and Nomenclature

Charles Lyell, in 1839, renamed his "younger Pliocene" the Pleistocene Epoch, the moniker meaning "most recent." The Tertiary-Quaternary boundary, or start of the Pleistocene, was defined in 1985 just above the top of magnetic polarity chronozone C2n (Olduvai) and the extinction level of the calcareous nannofossil *Discoaster brouweri* at the top of sapropel layer 'e,' Vrica section, Calabria, Italy, dated to 1.8 million years ago. The base of the Holocene, or Pleistocene-Holocene boundary, has an informal working definition of "exactly 10,000 Carbon-14 years ago" [which equals 11,500 calendar years ago, since a ^{14}C 'year' is not exactly equal to a calendar year], "at the end of the Younger Dryas cold spell." The present, of course, needs no definition. The Quaternary Period thus spans the most recent 1.8 million years of geologic time, and the Pleistocene has only a slightly shorter duration. I'm going to leave it at 1.8, or round it to 2, to save you fretting over a bunch of decimals.

Paleogeography and Paleoclimate

We've seen in the preceding chapters that the Earth had been inexorably headed towards the Ice Age for nearly 50 million years. Climates the globe over had been cooling—albeit somewhat erratically—since the mid-Eocene. The emergence of the Isthmus of Panama may have been the last straw that tipped our planet into a full-blown glacial age; one way or the other, the Pliocene closed with the Ice Ages just beginning, and the Pleistocene saw them develop in full glory.

Glaciers form when more snow falls in the winter than melts in the summer. The absolute amount of snow is relatively unimportant; it might be a dozen meters a season, but if it all melts in the summer, no glacier will form—or it could be only a couple of centimeters, but if summers are cold, an ice cap will slowly accumulate. Thus summer temperatures are frequently more of a controlling factor in ice sheet buildup than winter's nadir. Apparently, at high northern latitudes, a point was reached in the neighborhood of two million years ago where snowfall outpaced melting.

The great ice sheets have come and gone some seventeen times since the start of the Pleistocene. At their zenith, ice covered a third of the Earth's land surface, and icebergs choked half the world's oceans. Greenland became entirely ice-bound, and remains so today. In North America *sensu stricto*, the most extensively glaciated continent apart from Antarctica, the glaciers crept out like giant amoebas from two separate centers: one, in the northern Rocky Mountains, spawned the Cordilleran ice sheet, whereas the other, centered over Hudson Bay, gave birth to the Laurentide ice sheet. At glacial maxima, these two ice caps coalesced along the eastern front of the Canadian Rockies, and sea ice even reunited Greenland with continental North America; during interglacials, when the ice melted back, the Cordilleran and Laurentide sheets parted ways and opened an ice-free corridor into the midcontinent several thousand years prior to the full disappearance of the ice.

American mastodon Mammut americanum, provenance unknown. University of Michigan Exhibit Museum of Natural History, Ann Arbor.

When the ice sheets were at their maximum, so much water was tied up in them that sea levels worldwide dropped some 130 m (425 ft). Large expanses of what we today consider the continental shelves became dry land. This affected the Atlantic and Gulf coastlines more than the Pacific margin of North America, because in the east and south, the flatness of the terrain means that a small drop in sea level exposes far more new territory than it does in the west, where the coast is steep. Of the seventeen glaciations identified in ice cores from Greenland, four major ones have been named—the Nebraskan, Kansan, Illinoisan, and Wisconsinan (from oldest to youngest). Each of these saw the great ice sheets advance southward from Canada into the American Midwest, and each diverted rivers into the Mississippi that had formerly drained into the St. Lawrence or Hudson Bay. At its maximum, the southern edge of the ice formed a cliff 160 m (525 ft) high reaching to the 40th parallel, along a line drawn from modern Long Island to the Ohio and Missouri Rivers and on out to Seattle. The tremendous weight of the ice, as much as 3 km (2 miles) thick in some places, depressed the continental crust—forming Hudson Bay where there was previously dry land. The thickness of the ice can be estimated from the amount of crustal rebound that has occurred since its last retreat. It is a matter of speculation, in fact, whether Hudson Bay will rise above the waves before the next great ice sheet begins to cover it.

Why these episodic glacial ages? Two major factors have been identified that are necessary for the accumulation of continental ice. Plate tectonics must set the stage: there needs to be extensive land area at high latitudes, or any snow that falls will melt into the ocean. (And even if the sea surface freezes, it cannot support a thick ice sheet.) All of the past "ice house" episodes we've discussed (in the Late Precambrian, the Late Ordovician, the Late Devonian, and the Late Pennsylvanian-Early Permian) also occurred when there were continental land-masses in polar latitudes, particularly Gondwanaland. Since only 2/7 of the Earth's surface area is dry land, mere probability dictates that high latitudes are more often covered by open ocean than host to terra firma or to small, landlocked seas. Incoming solar radiation and outgoing heat loss balance at approximately 40° north and south latitude, so if the poles do not receive a constant heat influx from the lower latitudes—primarily through the circulation of ocean currents, but aided by atmospheric heat transport—they slowly cool off and eventually freeze. We have seen that there was a general cooling trend throughout most of the Tertiary, as Antarctica settled in over the South Pole and the Arctic Ocean became largely cut off from mixing with other waters. This configuration restricted the poleward transport of heat from the tropics, allowing the buildup of glaciers on Antarctica as far back as the Oligocene. By the Late Pliocene, glaciers had also begun to grow in the mountains of North America, although there were as yet only inceptive continental ice sheets.

Calcite in fossil whelk *Busycon* sp., Anastasia Formation, Indrio pit near Ft. Pierce, St. Lucie Co, Florida. Private collection.

So the drift of the continents puts in place a map which allows the polar latitudes to become intensely cold. Variations in astronomical parameters then control the waxing and waning of the glaciers. First, the Earth's axis is tilted with respect to the plane of its orbit around the sun: this is why we have seasons. The tilt is currently 23.5° (and decreasing), but varies between 21.8° and 24.4° with a period of approximately 41,000 years. The greater the tilt, the stronger the difference between the seasons. We have mentioned that it is not so much the coldness of the winters, but the coolness of the summers, that controls the growth of ice sheets—no matter how frigid the winter nadir, if summer melts all the snow that falls, no glaciers will build up. But if summertime temperatures are insufficient to melt winter snowfall, glaciers will form. Thus the strength of seasonality is a critical factor in the growth of ice sheets. It may be counterintuitive, but this means that winter cold during glacial periods may actually be less intense than it is during interglacials!

Another parameter which varies, this time with a periodicity of about 100,000 years, is the elongation of the Earth's orbit. Our planet does not revolve about the sun in a perfect circle—the orbit is instead an oval or ellipse, with the sun at one focus. The ellipticity of the orbit varies from only 3% (a near circle) to around 12% (a pronounced ellipse). This has no effect on the total amount of sunlight the Earth receives during the year, but again influences its distribution: the greater the ellipticity of the orbit, the greater the difference in the amount of sunlight received at perihelion (when Earth is closest to the sun) vs. the amount of insolation received at aphelion (when the Earth is farthest away). For the northern hemisphere (which currently contains the most land), when aphelion coincides with northern winter and perihelion with northern summer, the effects are additive and seasonality is enhanced; the opposite (which is occurring now) ameliorates the difference between summer and winter. This, therefore, also influences the growth and melting of ice sheets.

The last of the three astronomical factors contributing to the Ice Ages is what is called "precession of the equinoxes," with a periodicity of about 22,000 years, whereby the axis of the Earth repeatedly traces a conical path in the sky. The Earth's axis presently points to Polaris as the North Star, but in about 12,000 years, Vega will be the pole star. (One can compare this effect to the wobbling of a spinning top as it slows down.) Again, this has no effect on the total amount of solar radiation we receive during the year, but it determines where along Earth's orbital path summer and winter fall. In conjunction with the other two factors above, precession may attenuate or enhance the winter-summer temperature differences.

Sabertooth
Smilodon californicus, La Brea tar pits, Los Angeles, California. George C. Page Museum of La Brea Discoveries.

These factors interact in complex ways, first worked out mathematically in the early 20th century by the Serbian scientist Milutin Milankovitch (and commonly called 'Milankovitch cycles'). By sometimes offsetting each other, and other times working together, they produce variations in seasonality that control the waxing and waning of the ice sheets. The variable way in which the astronomical factors combine means that glacial-interglacial cycles are stronger and longer at some times than at others. And once continental ice has begun to form, a positive-feedback loop is initiated by the high albedo (reflectivity) of the white snow and ice: more snow/less dark-colored land = more heat lost to space, leading to further cooling of the polar regions and further growth of the ice. (You may recall some of this from our previous discussion of Precambrian Ice Ages.) Snow reflects 90% of the incoming sunlight, and absorbs only 10%; by contrast, forested land absorbs 90% and reflects 10%, and unforested land has median values. A restricted ocean such as the Arctic, which is capable of freezing over, is subject to this effect in spades, since open water absorbs even more of the incoming sunlight (some 97%) than land does. This feedback mechanism helps sustain an Ice Age once it has begun, and is responsible for the lower overall mean temperature during a glacial stage.

Sikrik
(Arctic ground squirrel) mummy, Spermophilus undulatus, frozen in its winter burrow. American Museum of Natural History, New York.

Glacials, on average, seem to last approximately 90,000 years, and interglacials usually about 8000 to 16,000 years. For most of Pleistocene time, therefore, the Earth was in the grip of glacial cold. From an historical standpoint, it also means that we should be nearing the end of the current interglacial, and heading inexorably into the next Ice Age. Whether this happens—or whether human-induced global warming (from the burning of rainforest vegetation and fossil fuels) offsets the other effects sufficiently to stave off the next ice advance—remains to be seen by future generations.

The Ice Ages were not only cooler overall—but less seasonal—than the interglacial climate familiar to us, but they were also a whole lot windier. Frigid katabatic (downhill) winds constantly blew off the ice caps, picking up copious dust and grit, especially from unvegetated, newly-deglaciated ground. Enormous deposits of loess (loose dust) were formed where these winds dropped their load. The strongest winds led to the formation of sand dunes, of which the Nebraska Sandhills are a remnant. The melting of the ice caused these winds to abate. Sediment cores from Lake Superior show a sequence of sand particles deposited by fierce winds and strong lake currents during the last glacial maximum, fining upward to clay as the winds died down with the onset of the Holocene interglacial. Ice cores from Greenland indicate that atmospheric dust was some dozen times what it is today. In many areas, especially on and close to the ice, snowfall was actually far less than it is now (only 1/3 to 1/2 as much), due to the combination of less evaporation from the colder seawater and to the extensive formation of sea ice which blocked easy access of maritime moisture to the continental interior. Antarctica and the North Slope of Alaska these days would qualify as deserts based on annual precipitation, but because of the cool summers, very little of the winter snow evaporates or melts. Thus the northern and interior regions of North America, during most of the Pleistocene, were cold and windswept, but with little snow cover.

The Western Interior of the United States, including the southwest, by contrast, received significantly more rainfall during glacial periods than it does today, because the ice caps diverted the jet stream (and hence main storm track) considerably southward. The copious rainfall swelled the Colorado River, with the result that most of the downcutting of the Grand Canyon occurred during Pleistocene time. Enormous lakes formed in the Great Basin: Lake Bonneville in Utah was the largest, rivalling modern Lake Michigan in size; Lake Lahontan in Nevada was almost half as big. Today, all that is left of Lake Bonneville is Great Salt Lake, the famous Bonneville salt flats, and the Sevier (usually dry) Lake. Some 15,000 years ago, Lake Bonneville's northern levee was breached, causing catastrophic flooding down the Snake River. Lake Lahontan is completely gone, though it succumbed much more quietly, to evaporation.

Sand dollars *Encope borealis*, El Golfo, Baja California, Mexico. Private collection.

Other bodies of water came into being during the early stages of an interglacial, when huge proglacial lakes formed from meltwater dammed up by the retreating ice front. Eventually the ice dam would break, resulting in giant waterfalls and rushing floodwaters. Glacial Lake Missoula in Montana repeatedly filled and drained in catastrophic floods, scouring the "channeled scablands" of the Columbia River plateau. Lake Agassiz—the largest proglacial lake by far, with four times the area of Lake Superior—at one time covered most of southern Manitoba, as well as portions of Minnesota, North Dakota, Saskatchewan, and Ontario; only Lake Winnipeg is left today. The rush of frigid meltwater into the Atlantic—via the Great Lakes and the St. Lawrence—when Lake Agassiz drained has been blamed for a thousand-year cold spell at the close of the Pleistocene called the Younger Dryas.

And while the legend of "Paul Bunyan's footprints" is beguiling, today's Great Lakes are actually just water-filled depressions scoured out of the soft Paleozoic bedrock by the mighty glaciers. The deepest, Lake Superior, lies atop a Precambrian rift, and is the largest freshwater lake in the world.

Life in the Pleistocene

Marine life throughout the Pleistocene was essentially modern, with the benthos (bottom fauna) dominated by hexacorals, clams, and snails; the plankton made up mainly of diatoms, dinoflagellates, and forams; and the open water the domain of sharks, ray-finned fish, and the great whales. One could have taken a scuba dive in a Pleistocene ocean and not noticed any significant differences from a dive today, as far as the character of the fauna is concerned. Specific areas, however, might have seemed quite different—with such events as the sighting of icebergs off the New England coast. Equatorial waters experienced far less temperature change than did polar ones as the ice sheets waxed and waned, although there was a contraction of the tropical reef zone by perhaps a few hundred miles during a glacial maximum. The most pronounced effect of the glaciations on sea life would have been the rise and fall of sea level itself, and the migration of the photic zone in concert with this. Fishermen today still dredge up mammoth and mastodon teeth from the seafloor of Georges Bank, revealing its history as dry land during the Ice Ages.

Terrestrial vegetation, too, consisted primarily of extant species, but in many places, the ecological associations have no modern analogue. In unglaciated far-northern lands, such as the interior of Alaska and the western Yukon, it is widely believed that the vegetation was a highly productive 'steppe-tundra' dominated by grasses, sedges, legumes, and dwarf trees such as willow, birch, and alder. A band of steppe-tundra perhaps a hundred miles wide also fringed the southern margin of the ice. This ecosystem supported vast herds of grazing mammals, such as mammoths, giant bison, caribou, horses, and musk oxen. The melting of the ice saw this vegetational association disappear and be replaced by the less productive Arctic tundra of today, with major effects on the fauna.

Yesterday's camel *Camelops hesternus*, Albuquerque gravel pit. New Mexico Museum of Natural History, Albuquerque.

The ice sheets pushed all the vegetational zones southward during their advance. Taiga (boreal coniferous forest), dominated by black spruce and tamarack and found mostly in Alaska and northern Canada today, covered the northeast, the Great Lakes region, and the eastern portion of the Great Plains. Abundant standing water, which sustains this type of forest and its associated boggy muskeg, resulted from reduced evaporation in the cool climate. Over most of this range, however, the woodland was less dense, and more open, than taiga today—more of a spruce parkland than a thick forest. This, again, is a biome that has no modern analogue. When the glaciers retreated, and the taiga advanced northward in their wake, various plant species dispersed in different directions, following their

Water beetles *Cybister* sp., Kern Co, California. Private collection.

preferred environments. The parkland association vanished, although the individual species did not. Other ecosystems, such as deciduous hardwood forest and prairie, closely resembled their modern counterparts but were found in more southerly latitudes than they are now.

The fauna of North America during these fascinating times was both familiar and exotic. Dire wolves and sabertooth cats are both Pleistocene icons. Many Tertiary denizens persisted into the Pleistocene, such as pronghorns, camels, horses, mastodons, giant beavers, and cougars. Other mammals, such as jaguars, bison, caribou, mammoths, and musk oxen immigrated from Eurasia during this time, by way of the Bering Land Bridge—which was hundreds of miles wide at sea level lowstands. In terms of the diversity of large mammals, Pleistocene North America was more comparable to modern Africa than to the North America of today.

Mammoths and mastodons particularly captivate the imagination and epitomize the Ice Age fauna. Three species of mammoth inhabited various regions of North America: the woolly mammoth, with its dense coat of dark hair, ranged across the northern steppe-tundra and stood 3 m (10 ft) tall at the shoulder; the Columbian mammoth dwelt on the prairie of the western Great Plains as well as across the southern half of the continent (as far south as Mexico and Central America), and was the largest mammoth, standing as much as 4.5 m (15 ft) high at the shoulder; Jefferson's mammoth was a somewhat smaller elephant with a wide distribution, and stood some 3.5 m (11 ft) tall. All mammoths were primarily grazers, as evidenced by their molars made up of many ridged plates—which looked remarkably like a stack of Fig Newtons seen on end (see p. 238). Such teeth enabled them to efficiently grind up abrasive, siliceous grasses and low-growing herbs. The American mastodon, by contrast, was a smaller (2.5 m/8 ft tall) browser that favored the spruce parklands of the northeast and Great Lakes regions. Its low, rounded, cusped teeth were adapted to a diet of mostly coniferous twigs and branches, as well as other leaves. Like the woolly mammoth, the mastodon sported a dense coat of long hair as protection against the cold.

Where Have All the Mammals Gone?

At the end of the last Ice Age, most of these megafauna vanished. Mammoths and mastodons, ground sloths, giant beavers, glyptodonts, and sabertooths went extinct; camels, cheetahs, and horses disappeared in North America, although some survive in South America and the Old World. With a few Canadian wood bison being the largest native mammals left, North America's fauna is severely depauperate compared with what it was throughout the Pleistocene and indeed, much of the Tertiary. What happened?

Two leading theories have been put forth to explain the demise of the Pleistocene megamammals: climate change and human hunting. There is no doubt whatsoever that climates changed radically with the onset of the Holocene interglacial. Whereas to us, it might seem as if warmer must be better, animals adapted to frigid conditions can actually have trouble adjusting to milder weather. It is the magnitude and rapidity of change, more than merely its character, that causes organisms difficulty in adapting. Indeed, of the extinctions that occurred *deep within* the span of the Pleistocene, most of them happened during interglacial periods, not during glacials. We have noted that while glacial conditions were cooler overall, the seasonality—or difference between the depths of winter and the height of summer—was actually less than we endure now. Along with this, glacial winters experienced less snowfall. Modern musk oxen can withstand temperatures down to -60°F (-50°C), but cannot tolerate snow cover of more than 10" (25 cm), because they cannot paw through it to reach their food. Flattened facets on the bottoms of mammoth tusks indicate that they were used (in part) for snow shoveling; perhaps these animals, too, were sensitive to the depth of snow. Increased winter snowfall during interglacials may thus have spelled doom for some animals which had weathered the glacial spells fine. Additionally, we have discussed how some Pleistocene vegetational assemblages have no modern analogue. The lush steppe-tundra ecosystem vanished altogether. Less nutritious grasses utilizing the C_4 photosynthetic pathway (which thrive under warm, dry conditions) edged out the more nutritious C_3 grasses (which do better in cool, moist climates) as temperatures rose; this would have decreased the carrying capacity of grasslands. Spruce parkland, the favored habitat of the mastodons, disappeared with the onset of the interglacial, to be replaced by a thicker spruce/muskeg taiga. Small animals, which need fewer resources (less food and a smaller home range) than large ones, perhaps could follow individual food plants or microhabitats as the parkland dwindled, while large animals like mastodons could no longer find sufficient resources to support a viable population.

Four-horned pronghorn *Stockoceros conklingi,* San Josecito Cave, Nuevo León, Mexico. Natural History Museum of Los Angeles County, Los Angeles, California.

However, there is equally no doubt that Paleoindians hunted at least a few of the Ice Age megamammals. There is some evidence of human immigrants from Asia as much as 50 thousand years ago (kya), in places such as Old Crow, Yukon Territory, and some 20 kya in caves in Peru. There is little, if any, certain evidence of human habitation of the midcontinent at this time, and a leading theory suggests that people may have followed a coastal route from Alaska south along the Pacific shore, settling near the ocean and travelling in boats. Many of these settlements perhaps are yet to be found on the continental shelf, drowned by the rise in sea level that accompanied the melting of the ice. Unequivocal evidence of widespread human habitation, however, does not appear until some 12 kya when an ice-free corridor emerged between the Cordilleran and Laurentide ice sheets and North America was peopled by what is known as the Clovis culture. These people, unlike their predecessors, hunted big game with finely-fashioned stone-tipped spears; Clovis points have been found associated with the remains at about 1/3 of mammoth fossil sites, and butcher-marks have been found on the bones of horses, camels, and bison, leaving no doubt that these were game animals. The argument can be made that earlier humans did not hunt these mammals to extinction because they did not possess the technology to do so, whereas the Clovis people did. However, this raises the question of why some of these game animals (like bison and caribou) escaped extermination.

This theory strikes a chord with us, because history is clear on what humans can and will do to other species if conservation is not practiced. The Maoris settled New Zealand around 1000 A.D., finding pristine islands with an unparalleled flightless-bird fauna. Within 600 years, they had hunted the ostrich-like moa to extinction, and the rats they brought with them had driven the tuatara (a lizard-like reptile) to the brink, where it survives today only on a couple of isolated, rat-free islets. Steller's sea cow, the only cold-water sirenian, was first reported by the Bering expedition in 1741. An easy source of food for sailors off seal hunting for 8 or 9 months a year, it was nearly gone when hunting was halted only 22 years later. Alas, although the last few sea cows were left alone, there were not enough of them to sustain a viable population, and by 1768 the species was extinct. Many more examples can be given: the dodo, the passenger pigeon, the Carolina parakeet, the Caribbean monk seal—all extinguished at the hand of man; North America's bison and pronghorns were barely saved, after initially large populations had been reduced to a tiny fraction of their former size by Europeans with guns. Even in the 'enlightened' 20th century, and sometimes even in the face of a hunting ban, mankind has driven the white rhino, the whooping crane, the Greenland right whale, the Asian arowana (dragon fish), and the California condor (to mention only a few) to the verge of extinction—and who knows how many rainforest species that have never even become known to science. Even if every last member of a species is not slaughtered, at some point a threshold is reached below which the population cannot recover. Breeding-age individuals may be too widely dispersed to find each other, viability may be decreased by inbreeding depression, or the population may be extinguished by a natural disaster. As previously noted, all cheetahs living today are as genetically similar as lab mice; they can accept skin grafts from one another. This degree of genetic homogeneity puts even a respectably-sized population at high risk of being wiped out by something like an epidemic disease.

We know it *can* happen, but *did* humans hunt the Ice Age mammals to extinction? Besides spearing, wasteful hunting practices certainly were employed—Head-Smashed-In Buffalo Jump in southern Alberta is only one of many such cliff-stampede sites known. But bison survived this degree of hunting, and flourished until Europeans with rifles came on the scene. Other mammals, though, may not have been able to endure even a small hunting pressure without succumbing. Populations were already stressed by the rapidly changing climate and vegetation, making them vulnerable. One study reported in the 1990s used computer modeling to estimate that Paleoindians killing only 2% of a mammoth population per year could drive the species below the sustainability threshold within a few centuries. Another study of growth rings in mastodon tusks—which, in a manner similar to tree rings, record good and bad years, and even births—concluded that mastodons in the Great Lakes region at the end of the last Ice Age were healthy and well-fed, but showed a birth rate similar to African elephants subjected to hunting. However, mastodon bones have rarely been found in association with human artifacts, and we know that the spruce parkland ecosystem on which these animals depended disappeared at the end of the Pleistocene. The "overkill" hypothesis also cannot account for the extinction of many birds (unlikely to have been game animals), or the reduction in size that many mammals experienced at the end of the Pleistocene—things more easily explained by climate change. And if you've been paying attention, you've noticed that numerous mammalian extinctions have been part of the North American faunal landscape since the Miocene: witness the reduction in horse species diversity from a dozen (then) to a single Pleistocene species, and a parallel trend amongst the pronghorns. Clearly, humans cannot be blamed for this.

The truth is that we will never know for sure, but in all likelihood the end-Pleistocene megafaunal extinctions were brought about by the combination of changing environments and human hunting. Climate change stressed these animal populations, but they had weathered interglacial times before, until a ferocious new predator appeared on the scene. And while this predator himself likely did not hunt the other predators (like sabertooths and dire wolves), he probably aided in bringing about their demise by contributing to the disappearance of their prey animals.

Columbian mammoth *Mammuthus columbi*, provenance unknown, Burpee Museum of Natural History, Rockford, Illinois.

Where Pleistocene Rocks Are Found

Ok, enough for the guilt trip. Where in North America are Pleistocene sediments found? The short answer is: just about everywhere. Most such deposits are surficial—unconsolidated sand, gravel, and dirt lying atop older bedrock—because little time has passed for lithification. Many do not even have formal formation names. These deposits are ubiquitous in formerly-glaciated areas, meaning almost all of Canada and the northern tier of the

Mammoth
tooth found by miners dredging a river near
Fairbanks, Alaska. Private collection.

United States. Cape Cod and Long Island are glacial moraines—sediments dropped at the margin of the ice sheets when their advance stagnated and their retreat began. The Midwestern states and provinces owe much of their topography to the moraines, eskers, drumlins, and kettle-hole lakes left by the glaciers. And the U-shaped valleys of the northern Rocky Mountains were sculpted by alpine glaciers during the Pleistocene; gravel bars—end moraines—frequently indicate the maximum extent of the ice.

The vanished lakes of the western states have also left dry beds of Pleistocene sediments, and much of the intermontane alluvium in the basin and range province was deposited during this timeframe. In short, a great deal of the "plain old dirt" around us is actually Pleistocene rock-in-the-making.

Where Pleistocene Fossils Are Found

Sometimes bones are found in glacial deposits—such as many of Michigan's mastodons—but lots of the best Pleistocene fossil sites are scattered concentrations in such localities as sinkholes, caves, and fissure fills. Because of this, it is impossible to paint a broad picture of where Pleistocene fossils can be found, but we can cite a smattering of individual localities scattered across North America.

First, let us enumerate some named formations that are known to be productive. Alaska's Cape Deceit Formation yields fossils of deer, caribou, pikas, lemmings, and voles. The Gubik Formation of Alaska's North Slope is full of petrified wood and clam shells. Horses, mammoths, and prairie dogs are found in Alberta's Hand Hills Conglomerate. The American Falls Formation, exposed at American Falls Reservoir in Idaho, is a good place to look for the bones of sloths, mammoths, bison, lions, horses, and camels—and new bones are exposed each year as the water level rises and falls. Nebraska's Grand Island Formation produces muskrats, wolves, horses, peccaries, and sabertooths. The McPherson Formation of Kansas contains horses, molluscs, fish, frogs, salamanders, snakes, turtles, and birds. The Seymour Formation of Texas has yielded the remains of mammoths and tortoises. South Texas' Beaumont Formation produces fossil mammals such as horses, mammoths, and camels. Florida's Melbourne Formation produces the bones of larger animals such as armadillos, bears, capybaras, beavers, tapirs, bison, mammoths, and horses. Mexico's Tacubaya Formation (in the state of Aguascalientes) produces the bones of horses, camels, pronghorns, giant bison, and the American lion. Also south of the border, in the state of Mexico, the Becerra Superior Formation yields fossils of sloths, giant armadillos, and bighorn sheep.

As aforementioned, most of the best Ice Age fossils come from localized pockets of sediment, rather than widespread formations. Port Kentucky Cave in Pennsylvania is a fissure fill in which the bones of sabertooths, black bears, and sloths have been found. Cumberland Cave in Maryland contains the best Early Pleistocene fauna in eastern North America, with dogs, marmots, mice, bears, peccaries, weasels, porcupines, muskrats, and voles. Most Pleistocene fossils found in Florida come from the state's rivers, including the Suwanee, Withlacoochee, and Caloosahatchee, which produce disarticulated bones of mastodons, manatees, giant armadillos, sloths, and tapirs, among others. Ice age fissure fillings in the Early Tertiary Ocala Limestone of Florida and Georgia are replete with fossils of small animals such as mice, bats, amphibians, box turtles, and birds. The same state's Devil's Den sinkhole yields sloths, dire wolves, bats, mammoths, sabertooths, horses, peccaries, muskrats, voles, and fish. The salt dome of Avery Island, Louisiana attracted a variety of animals as a mineral lick, and the fossils of mammoths, mastodons, sloths, horses, and bison can be found there. Trollinger Spring, Missouri is the site of many finds of mastodon, horse, and musk ox bones. Friesenhahn Cave in Texas was a sabertooth den where young mammoth kills were cached and eaten; 13 cubs and 20 adults of the scimitar cat *Homotherium*, and some 441 juvenile mammoth molars, have been recovered from the cave. At Blackwater Draw, New Mexico human artifacts have been found in association with mammoth kills, as well as the bones of dire wolves, sabertooths, peccaries, camels, bison, and horses. It is likely that hunters waited at the spring and surprised mammoths that came down to drink there. San Josecito Cave in the central mountains of Nuevo León, Mexico has produced fossils of mountain goats, deer, dire wolves, mountain lions, and bears. Eight skeletons of the ground sloth *Eremotherium* have been found at El Hatillo, Panama. Numerous caves in the desert southwest have yielded the bones or mummies of ground sloths, along with valuable deposits of dried dung that can be analyzed for clues to the sloths' diet. Rampart Cave in northwestern Arizona used to contain the largest known and best-preserved deposit of ground sloth dung, until a fire started by vandals destroyed most of it in 1976. More sloth dung, along with bones, hide, and hair, is found in Gypsum Cave east of Las Vegas. A small-mammal fauna from the end of the Ice Age comes from Snake Creek Burial Cave in Nevada. Kokoweef Cave, California preserves the remains of Pleistocene gastropods, amphibians, reptiles, birds, pikas, and ground squirrels. Fossil Lake, Oregon is the most important Late Pleistocene fossil site in the Pacific Northwest, and preserves a menagerie of sabertooths, ground sloths, camels, dire wolves, mammoths, horses, rodents, birds, and fish. Moonshiner Cave

Giant beaver *Castoroides*, peat bog, midwest North America. North American Museum of Ancient Life, Lehi, Utah.

in Idaho has a preponderance of carnivore bones, especially wolverines and weasels. The richest Pleistocene vertebrate assemblage known from the Great Plains comes from bluffs along the South Saskatchewan River near Medicine Hat, Alberta, and includes fossils of bison, American lions, horses, camels, caribou, and mammoths.

In the Arctic regions of Canada and Alaska, woolly mammoth carcasses periodically melt out of the permafrost, and have been known to native peoples for millennia. Tusk ivory from such animals has traditionally been collected, traded, and carved. Ancient peoples used to think the behemoths actually lived underground (the generic name for the mastodon, *Mammut*, means "earth burrower"), but it is now known that the remains are those of unfortunate victims that became mired in mud or fell into ice crevasses and have been frozen since the Pleistocene. These finds preserve not only the bones, but also muscle, skin, hair, and sometimes stomach contents of the animals. In 1979 the mummy of a steppe bison was discovered during streambed mining in Alaska, and nicknamed "Blue Babe" after the azure mineral vivianite that impregnates the carcass, and Paul Bunyan's legendary ox companion, Babe. This fascinating mummy is now on display at the University of Alaska Museum in Fairbanks.

One might easily get the impression that all Pleistocene mammals were large. True, many were, but then as now, a host of smaller mammals lived in the shadow of the giants. Porcupine Cave in central Colorado's mountains preserves the best high-altitude Pleistocene fauna in North America. Most of the bones there are small ones collected and cached by packrats building their nests. Mice, voles, rabbits, prairie dogs, and even a black-footed ferret have been recovered from the cave. The stratified sediments go back a million years, and preserve the comings and goings of several glacial cycles—and are particularly important in studies of changes in the montane fauna over time.

Where Can Pleistocene Fossils Be Seen in the Field?

For a snapshot of North America's Ice Age fauna, there is no better place to visit than the Rancho La Brea tar pits in downtown Los Angeles, California. These asphalt seeps came into existence some 40,000 years ago and remain active today. Crude oil seeps to the surface through fissures in the bedrock; the volatile fraction evaporates, leaving behind thick pools of asphalt that make molasses seem as thin as water. Animals became trapped in the pools, died, and were preserved by the tar. (At the George C. Page Museum of La Brea Discoveries, there is a fascinating exhibit where you can try to pull a metal rod the size of a mammoth leg out of the asphalt—you are lucky to move it a few inches, and it really gives you a good idea of how hopelessly mired an animal could become).

Fossils recovered from the tar paint a vivid portrait of the Ice Age fauna: bones of mammoths, bison, giant ground sloths, pronghorns, horses, camels, short-faced bears, sabertooth cats, dire wolves, coyotes, and large predatory birds called teratorns have all been found there, along with plant remains, smaller vertebrates, snails, and insects—all of which are useful in reconstructing the paleoecology of the area. Interestingly, carnivores—dire wolves in particular, but also sabertooths and coyotes—vastly outnumber herbivores at La Brea. In modern ecosystems, herbivorous mammals generally outnumber carnivores by about 10:1. The predominance of carnivores in the tar pits is attributed to predators and scavengers having been attracted to a dead or dying prey animal trapped by the tar, only to become mired themselves. However, far from being a usual occurrence, the entrapment of only an average of one animal per year would account for the vast number of bones (some three million) that have been found there.

For a fossil assemblage similar to that found at the Tar Pits but in a rustic, rural setting, check out the Manix Lake Beds in the Mojave Desert near Barstow, California. This is a BLM "area of critical environmental concern" so no collecting is allowed, and there are no developed facilities, but you may spot bones of fish, turtles, or any of the mammals mentioned at La Brea. The fossils are found in a dry channel of the Mojave River in Afton Canyon, and accumulated between 450,000 and 19,000 years ago.

Two other natural traps are also among the most famous Pleistocene fossil localities in North America. At Big Bone Lick near Union, Kentucky, a bog associated with a salt lick preserved unfortunate animals which were attracted by the salt and became stuck in the mud. Mammoth, mastodon, bison, horse, musk ox, and ground sloth bones are amongst the menagerie found there. Now a state park, this locality boasts the distinction of being one of the first-discovered fossil deposits in the United States. The salt licks were well known to the Delaware and Shawnee tribes. Fossil bones were initially reported by an Indian trader in 1744; Thomas Jefferson sent William Clark (of Lewis and Clark) to collect bones there in 1807, and some of the specimens are still on display at Jefferson's Virginia home, Monticello. If you visit the

Dire wolf Canis dirus, La Brea tar pits, Los Angeles, California. Sternberg Museum of Natural History, Hays, Kansas.

park, you can take a trail to the original salt spring, and view an outdoor "bog diorama" featuring reconstructions and skeletons of some of the animals found there in a recreated Ice Age habitat.

A more recent discovery was made at Hot Springs, South Dakota in 1974 during bulldozing for a housing development. A cache of mammoth bones was uncovered, work was halted, and scientific excavations were begun. Since that time the bones of 49 Columbian mammoths, mostly young males, have been recovered from what had been a Pleistocene sinkhole. Apparently the mammoths came down to the pool to drink, fell in, and were unable to climb out the steep, slippery sides. Today the Mammoth Site is covered with a museum building where visitors can view the ongoing excavations. While the mammoth fossils are the most spectacular ones in the menagerie, the site has also yielded the remains of bony fish, toads, birds, rabbits, squirrels, prairie dogs, wolves, coyotes, camels, pronghorns, and the giant short-faced bear.

With the passing of the Pleistocene, North America, like the Old World, came to be dominated by a single animal—*Homo sapiens*. Our species changed the continent forever, harnessing the environment to our purposes and probably contributing to the extinction of the Ice Age megafauna. But for our appearance here, the Holocene would be just another mundane interglacial in a series of who knows how many advances and retreats of the ice. What has our continent seen in the eleven-and-a-half millennia since the collapse of the last great glaciers? What lessons can we learn from North America's past, and what story line are we writing now? These questions will form the core of our final chapter, much as the Canadian Shield has formed the core of North America since we embarked upon our journey.

Chapter Eighteen
The Holocene

The first snow dusts the Rockies of southern Colorado two thousand years ago. A bison herd grazes in the valley, three mule deer enjoy the autumn warmth, and a coyote ambles by, heedless of the glowing gold of gaillardias. A bald eagle soars overhead and an Arapahoe village nestles in the lee of a sandstone massif; blazing sumacs herald the change of seasons.

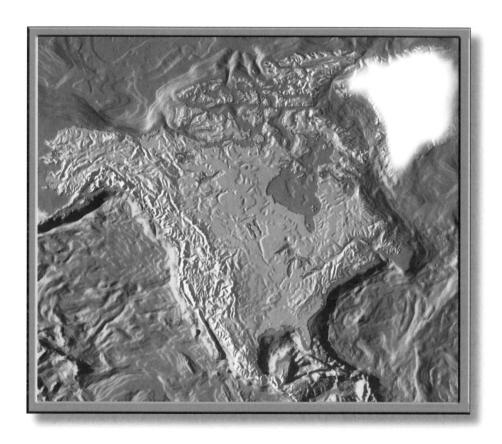

Holocene North America

he great ice sheets were nearly gone from the mainland of North America: by 11,500 years ago, we had entered an interglacial, one of the warm spells that periodically characterize about 10% of "Ice Age" time. Were it not for the rise of civilizations, this interglacial would be unremarkable. Instead, our continent entered the Holocene with an impoverished fauna, and the changes North America would see in the short span of a few millennia were to be some of the most far-reaching since its initial assembly during the Precambrian.

Definition and Nomenclature

Charles Lyell originally used the term 'Recent' for "the time that has elapsed since the Earth has been tenanted by Man," which consisted of the last fragment of Quaternary time. In 1867 M. Gervais proposed the name Holocene, meaning "fully recent," for this Epoch—tellingly noted as "the post-diluvial deposits approximately corresponding to the post-glacial period." Ice Ages were accepted, after having been identified by Louis Agassiz in the mid-nineteenth century from his studies in the Alps, but Biblical dogma was not yet divorced from geology. The International Geological Congress ratified the Holocene moniker in 1885.

The base of the Holocene, or Pleistocene-Holocene boundary, has an ICS informal working definition of "exactly 10,000 carbon-14 'years' ago (~11,500 calendar years B.P.) at the end of the Younger Dryas cold spell." For the purposes of "B.P." (Before Present), the "present" is set at A.D. 1950. Thus, if you care to, you can calculate in *exact* calendar years just how long ago the start of the Holocene was: at this writing, it is 11,557 years ago. By the time you read this, it will be 11,558 or so. For discussion purposes, I'll stick with 11,500 as close enough. The Holocene is, therefore, by far the shortest epoch we have examined, at eleven-and-a-half millennia long.

Geography and Climate

The geography of North America has changed very little since the Holocene began. Plate tectonic movement has been less than a kilometer (.6 mile). Topographic changes have largely been related to drainage patterns and rebound of the crust in the wake of the melting ice. The ice sheets were responsible for diverting the Missouri River drainage from Hudson Bay into the Mississippi, where it flows today. The Great Lakes filled with meltwater as the ice retreated, and the great proglacial lakes drained away. Isostatic rebound of the crust in northern Canada has elevated the Hudson Bay region several hundred meters (~1000 ft), and the glacial meltwater has raised sea levels about 180 m (600 ft), flooding the continental shelves. Crustal rebound continues today, albeit at a slower pace.

Although the major retreat of the ice cap north of the Great Lakes region had occurred by the onset of the Holocene, the last remnants of the Laurentide (Hudson Bay) ice sheet in Labrador did not melt until some 6500 years ago. The Cordilleran (mountain) ice cap fragmented and disappeared, leaving disconnected alpine glaciers in its wake, mostly in Alaska and Canada. The maximum summer insolation was reached about 10 kya; since then, incoming solar radiation has been slowly waning. Temperature depression attributed to thermal inertia of the remnant ice caused the maximum climatic warmth—termed the hypsithermal—to be reached about 5500 years ago in the eastern half of the continent, approximately coinciding with the start of the Bronze Age in the Old World. Maximum temperatures were reached earlier (around 9 kya) in western North America, and were some 2° - 3°C (4° - 5°F) warmer than today. Stranded driftwood on Ellesmere Island attests to a mostly ice-free Arctic Ocean between 6 and 4 kya. Subsequently, the slow descent into the next Ice Age has begun.

Five thousand year old packrat midden, Chaco Canyon, New Mexico. These rodents collect seeds and small objects to make their nests and cement them together with urine. New Mexico Museum of Natural History, Albuquerque.

The cooling has been gradual and not entirely smooth. Montane glaciers in the North American west began to expand some 5 kya, with other episodes of growth around 3 kya and 400 years ago. The last of these 'Neoglaciations' is called the Little Ice Age, and is particularly well-known since it falls within the realm of relatively recent historic times. Although accurate temperature records have only been kept since the mid-1800s, it has been possible to infer temperatures back several centuries by looking at such things as the records of European vineyard harvests. (When you could grow grapes in England, you know it was warmer than it is now, and by how much.) By this we know that the Little Ice Age followed a modest warming in Late Medieval times and lasted about 400 years, from the 14th to the 18th century. Since that time, Earth's climate has been warming again—a matter we will discuss in more detail in just a bit.

Life in the Holocene

At the hypsithermal, the northern tree line (the latitudinal limit at which trees can grow) was several hundred kilometers further north than it is now—as shown by subfossil forests growing in areas that support only tundra today. Tree line in the mountains was at a higher altitude as well. Prairies also expanded their northern and eastern reach by about 100 km (60 miles) before the transition to boreal and mixed hardwood forest occurred. Beavers lived on the Seward Peninsula in Alaska, far north of their present range.

The megafauna were gone. Bison were the largest animals left, and they filled the prairies in vast herds stretching from horizon to horizon. The camels had emigrated to South America and Asia, the horses to Asia and Africa. Pronghorns kept mainly to the sagebrush steppe of the Great Basin and surrounds. Deer were ubiquitous, ranging from coast to coast. Musk oxen and caribou inhabited the far north, and moose ranged across Canada and the northern tier of states. Wolves, foxes, coyotes, cougars, and black-footed ferrets preyed on the large herbivores or the small rodents. Innumerable prairie dog towns dotted the High Plains. Eagles soared.

Though our discussion of Paleoindians' probable role in the end-Pleistocene extinctions has laid bare the myth of the "noble savage," always living in harmony and equilibrium with the environment, nothing had prepared the fauna of North America for the coming of Europeans. The slaughter of the herds of bison—for trophies, tongues, and hides, the meat left to rot in the baking sun—need not be recounted here. Smallpox, war, and confinement to reservations did the same for natives of our own species. Shotgunning of the passenger pigeon reduced flocks numbering in the billions, which used to darken the sky, to a lone female, Martha, who died in the Cincinnati Zoo in 1914. But guns didn't do it all. The vast majority of species which have been extinguished at the hand of Man—here and elsewhere—succumbed to our usurpation of their habitat. The coyote is one of very few native mammals which have expanded their range since the coming of the white man—a mute testimony to the adaptability of this highly intelligent species.

Forests were razed for their lumber and to clear land for farms. The Great Plains yielded to the plough; in their western reaches, Pleistocene water from the Ogallala aquifer is still needed to irrigate the farms, and it is not being replenished. Each passing year sees more concrete laid, more steel reaching towards the sky. To make room for us, the other inhabitants of our continent have been squeezed out of the way. Not all have survived; the whooping crane and the California condor have been lucky. Will those of us who are here—now—choose to share our continent with future generations, not only of humanity, but of all species?

The Global Challenge

On our journey through North America's past, we've seen the climate change time and again, sometimes hot and dry (like in the Triassic), sometimes cool and wet (like in the Pennsylvanian), sometimes swinging erratically (like in the Pleistocene). There is no such thing as a "normal" climate—just the temporary result of complex interactions between the dance of the continents, the flow of the ocean currents, the composition of the atmosphere, the albedo of the land, and a multitude of other factors. But these changes typically operate slower than molasses in January compared to a human lifespan. "Normal," as we have come to think of it—and built our civilizations around—is a Pleistocene interglacial. We've seen that our interglacial has already peaked and the descent into the next Ice Age begun. But the Holocene is not a typical interglacial; we are here now, with our cities and our technologies. We have the ability to alter our climate (if not to harness it), and most scientists (if not most politicians) agree that we are already doing so.

We all remember carbon dioxide, and its role as a "greenhouse gas." Measurements from air bubbles trapped in Antarctic ice cores indicate that atmospheric CO_2 was about 284 ppm at the beginning of the Industrial Revolution. In less than two centuries it has risen to 383 ppm, and various projections call for the level to double or quadruple within the next century and a half. Most of this is attributable to our burning of coal, gas, and oil to heat our homes, fuel our cars, and power our industry. At the same time, we are cutting down forests at an alarming rate, particularly in tropical regions. Those forests are massive carbon sinks; remember what happened in the Devonian as trees spread across the landscape? The net result is an expected global temperature rise of 2° - 3°C (4° - 5°F) within the next fifty years—a return to the Holocene hypsithermal. This is on top of the +0.6°C (1°F) warming that already took place during the 20th century. And that's only the beginning. If emissions are not checked, and atmospheric CO_2 continues to rise at the present rate, we could be looking at global warming on the order of 8°C (14°F) within a hundred years—comparable to a return to the climate of the Late Cretaceous. Unzip your parka, and get out some sandals and a straw hat...

So what? I remember the good ol' Cretaceous! Warm and balmy everywhere, perpetual summer...it'd be like a tropical vacation all year 'round! Those doomsayers don't know what they're talkin' about...the planet wouldn't fry...we've 'been there' before!

Wait a minute. You've got a definite point. The planet wouldn't fry. But let's examine the situation a little more closely. Take a look at that Cretaceous map on page 150. Where do you live? My home in Boulder, Colorado looks like it would be seaside real estate. I'm not sure I'd like that; I moved here to get away from the clouds and humidity of parts east. Uh-oh. Uncle Greg and Aunt Sandy live in Florida—well, they used to. I wonder where they'll go now. Looks like there might be some real estate opportunities in Greenland. Karen and Elizabeth are probably both looking at a move, too, what with the look of that new California coastline. Linda and Rockey will be all right—they'll be close to the beach, now, right there at their place in Albuquerque. But both being born and raised in the desert, I doubt if they even own bathing suits. Allen's home outside Chicago, and Marc's in southern Ohio, will be safe. Looks like they'll be on a separate continent from me, though—won't be able to drive back east next time I visit. Sally's awfully close to Puget Sound, but she's been talking about wanting to retire to the Southwest anyway...

As the above thought experiment shows, global warming has consequences far beyond just shorter shirtsleeves. Sea levels are unusually low right now because of all the water locked up in the Greenland and Antarctic ice caps. If those were to melt—to any substantial degree, even if not totally—we would see major flooding of low-lying coastal areas. Sea level is already up some 15 cm (6 in) in the last century, and is currently rising at 3 mm/yr. Most of the in-

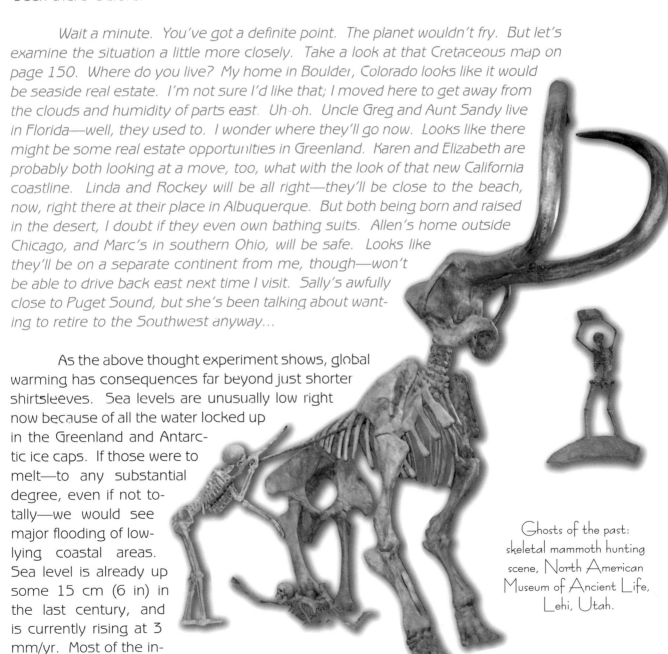

Ghosts of the past: skeletal mammoth hunting scene, North American Museum of Ancient Life, Lehi, Utah.

crease so far is from thermal expansion of seawater as temperatures rise, and thus doesn't even include the additional volume of glacial meltwater. A 3.5 m (11 ft) sea level rise would flood most of south Florida (including Tampa/St. Pete, Miami, and West Palm Beach) and southern Louisiana (including Baton Rouge and New Orleans). This is within the realm of projections based solely on the expected increase in atmospheric carbon dioxide. Air temperatures along the southeast coast of Greenland have increased 3°C (5°F) in the last two decades, and the Greenland ice cap is already melting much faster than scientists anticipated. The EPA-projected 60 cm (2 ft) sea level rise in the next century may well be quite on the low side. Several computer models predict a 7 m (23 ft) sea level rise within a millennium or less. Of course, the Western Interior Seaway wouldn't reappear immediately, so long as mantle upwelling keeps the western third of the continent elevated. However, the Cretaceous is the nearest analogue we have for global warming of this magnitude.

Fossils in the making: cow skull and yucca, Eldorado Springs, Colorado. If undisturbed and eventually buried, in a million years or so, these may become fossils.

But wait—there's more! Recall the end-Permian and end-Paleocene temperature spikes? Does the term 'methane hydrate' ring a bell? Holy moly, we've got those to worry about too. Global warming doesn't stop with carbon dioxide if it increases the temperature of the oceans enough to cause them to 'burp' out a substantial amount of that seafloor methane—lying there like a land mine, just waiting to be tripped. Remember that methane is a much more powerful greenhouse gas than carbon dioxide. Suddenly complete melting of both the Greenland and Antarctic ice caps is well within the realm of reality. That would discharge enough meltwater into the oceans to cause a sea level rise of 80 meters (263 ft). Turn around, and New York and Houston are submerged now. But cheer up; all those politicians in Washington, D. C. who wouldn't listen back in the early 21st century are drowning too...

Some places would benefit; Greenland and the Canadian Arctic would become habitable. But it would have major consequences on the human scale. Warming would cause the grasslands—the Grain Belt, the Breadbasket of America—to migrate north into Canada. Nothing against all you Canucks, we love ya, but this is a serious economic power shift. All of a sudden, we Americans live "south of the border." And Mexico? Probably even hotter, drier, and poorer than it is now.

The problem with both the melting of icecaps and the release of seafloor methane hydrates is that they are threshold phenomena. They involve physical state changes, which happen suddenly when the transition temperature (melting or sublimation) is reached. A glacier can surge to the sea on a lubricating layer of meltwater a hundred times faster than the ice itself can flow. Release of seafloor methane can set up a positive-feedback loop leading to runaway global warming—one that, once started, might be impossible to stop. Climate change isn't slower than molasses in January any more; even the end of the Pleistocene seems tame by comparison. Humanity will be hard-pressed to adapt, but sea level is unlikely to rise faster than we can run out of the way. Other species, however—plant or animal—may not be so lucky. Will scholars a thousand years from now be analyzing the Late Holocene mass extinction?

We are fortunate that the United Nations has taken the lead in this matter. In 1997, an international panel of experts drafted the Kyoto Protocol, an amendment to the United Nations Framework Convention on Climate Change. This voluntary, binding agreement seeks to reduce the CO_2 and other greenhouse-gas emissions of signatory nations (by variable amounts, depending on their current output and circumstances) in hopes of stabilizing them at levels that will prevent dangerous anthropogenic interference with the climate. Most civilized nations (169 as of December, 2006) have chosen to ratify the Kyoto Protocol, and it was officially entered into force on February 16, 2005. Sad to say, my government is not among them—and I live in the country with the highest total CO_2 emissions in the world, so our participation in this effort is critical. Germany and Britain have already reduced their CO_2 to 17% and 14% below 1990 levels, respectively, demonstrating that committed, industrialized nations *can* meet their targets without going bankrupt. Last year the City of Boulder voted to adhere to the Kyoto restrictions ourselves, regardless of the inaction of the federal government. Many other cities, several northeastern states, and the states of California and Oregon all also have some form of action plan in place. There is a small tax on my monthly gas bill to support compliance—the price for my affirmative vote. I'm glad to pay it. Only with the support of individual citizens can these problems be addressed.

A curious juxtaposition: a sculptural rendition of a mammoth trapped in the La Brea tar pits, silhouetted against the skyline of downtown Los Angeles.

We can all do our share. I don't live in a solar-heated cabin in the mountains, eating nothing but yogurt and granola. I do recycle. I keep the thermostat low in the winter and off in the summer, use lots of blankets and cool my house with fans. I buy wind energy to cover my electric usage. But I don't bicycle everywhere. Not everyone's health allows it. I do drive a small car. You have to choose what works for you. Not all communities have recycling programs, and you'd be crazy to forego the a/c in a Phoenix summer. But bit by bit, starting at home, working together, we can avert disaster—like Kipling's Old Man Kangaroo, we have to.

We've followed North America's journey for some four billion years now. We've seen our continent assemble from islands of scum; rise above the waves, only to be flooded again; turn from bare, windswept crags to a lush, velvety green; and support an ever-changing menagerie of animal life—some fierce and bizarre, some bold and majestic, some cuddly and tame. Through it all, our land has stood proud against the sunset and been home to us all. Now we are facing new challenges, as humans, like the humble cyanobacteria of the Precambrian, have evolved the ability to wreak havoc with our environment. Can we learn from the past, or is our fate to be the same as the dinosaurs'—extinction at the hand of environmental destruction?

Glossary

A

abiotic - inorganic, not living

abscission - the act of a leaf being naturally shed from a tree

aerobic - oxygen-using or -requiring

agnathan - without jaws (usually referring to fish)

albedo - reflectivity, or measure of the amount of sunlight reflected back into space

amnion (amniotic sac) - a membrane surrounding the developing embryo found only in higher vertebrates

amniote - an animal possessing an amnion in early development

anaerobic - without oxygen (frequently referring to metabolism)

anapsid - tetrapod with no postorbital fenestrae (holes in the skull behind the eye)

angiosperms - flowering plants and trees

anoxic - lacking in oxygen (frequently referring to the environment)

arborescent - tree-like

articulated - with all the bones in life position

artiodactyl - even-toed hoofed mammal

B

basalt - a type of igneous (volcanic or plutonic) rock with high iron content

benthic - bottom-dwelling

biostratigraphy - the science of dividing and correlating rock layers based on the fossils they contain

bipedal - walking or standing on two legs

boreal - northern

C

caliche - crusts of calcium carbonate, indicating a high rate of evaporation

carnivore - meat-eater

chloroplast - plant organelle containing chlorophyll for photosynthesis

cladistics - also called 'phylogenetic systematics,' a method of classifying plant or animal taxa based on shared derived characteristics, very popular nowadays

cursoriality - fast running

D

deciduous - shed annually; usually refers to tree leaves, but can also refer to antlers or horn sheaths

dentition - teeth

desiccation - drying out

detritivore - an animal that eats detritus, i.e. small bits of decaying plant and animal matter, usually found in the soil or on the sea bottom diapsid - tetrapod with two postorbital fenestrae (holes in the skull behind the eye)

drumlin - glacially-derived oblong hillock

E

ectothermic - cold-blooded

endothermic - warm-blooded

eolian - wind-deposited

epicontinental (seas) - shallow seas that are basically flooded continental areas

epiphyte - a plant that grows attached to a tree; e.g., many orchids and bromeliads

esker - sinuous gravel formation left by river running beneath a glacier

eukaryote - a cell with a nucleus

extant - still living

extinct - no longer living, as in an entire species

F

femur - thigh bone

flowstone - limestone cave formation that seems to flow like a curtain

folivory - leaf-eating

foraminifera (forams) - tiny, shelled marine amoebas

G

gallery forest - forest lining a river or stream

gnathostome - fish with jaws

gymnosperms - non-flowering seed plants which commonly (though not always) bear their seeds in cones

H

herbivore - an animal that eats plants

holarctic - referring to the modern circum-arctic continents of North America, Europe, and Asia

homeobox (Hox) genes - a class of genes which control developmental processes, rather than the synthesis of proteins

hypoxic - low in oxygen

hypsithermal - time of maximum warmth during the Holocene

I

igneous rock - rock that was once molten deep within the Earth, and has solidified either underground or upon eruption from a volcano or fissure

in situ - in its natural place

inertial homeotherm - an ectothermic animal that can maintain a fairly constant body temperature by virtue of its large size and thermal mass

insectivore - an animal that eats primarily insects

invertebrate - animal without a backbone

K

kettle-hole - lake formed by hole left when a block of ice was stranded and left behind as a glacier melted back

L

lagerstätte - an exceedingly rich or well-preserved fossil deposit

Lazarus taxon - a group which disappears from the fossil record for a period of time, only to reappear later

M

metamorphic rock - rock (usually of sedimentary origin) that has been reheated and recrystallized but at a temperature below its melting point

metazoan - animal

mitochondrion - organelle responsible for energy generation; "powerhouse of the cell"

molecular clock - a method of estimating the time since two species shared a common ancestor, based on the amount of genetic difference between their DNA

monophyletic - a group (called a clade) which contains all descendants of a common ancestor

monsoon - strong seasonal rain generated by temperature differences between the ocean and a large continental landmass

moraine - sand, gravel, etc. dropped by a glacier when it begins to melt back

morphology - outward appearance, e.g. shape, size, etc.

N

nonvascular plants - plants lacking a water circulatory system, relying instead on osmosis for water transport; e.g. mosses, liverworts

O

omnivore - an animal that eats a varied diet; humans are omnivores

ornithischian - "bird-hipped" dinosaur (with posteriorly-pointing pubic bone)

orogeny - an episode of mountain-building

P

pelvis - hip bones

perissodactyl - odd-toed hoofed mammal

photic zone - the water depth to which sunlight reaches

photosynthesis - the process by which plants use sunlight to make sugars for food

piscivore - an animal which eats fish

planktonic - free-floating

proglacial lake - a lake that forms because it is dammed up by a glacier

prokaryote - a cell without a nucleus

provenance - geographical derivation or whereabouts

pteridosperms - "seed ferns"

Q

quadrupedal - walking or standing on four legs

R

radiometric dating - dating of rocks based on the decay of radioactive minerals they contain

rhizome - creeping stem, frequently anchored to the ground by rootlets

S

saurischian - "lizard-hipped" dinosaur (with anteriorly-pointing pubic bone)

sedimentary rock - rock formed from the settling and compaction of small particles of mud, silt, sand, shells, etc.; e.g. sandstone, limestone, shale

sensu lato - in the broad sense

sensu stricto - in the strict or narrow sense

sessile - unable to move around

symbionts - two organisms living together, frequently in a mutually beneficial relationship; e.g. the fungus and the alga that together make up a lichen

synapsid - tetrapod with one postorbital fenestra (hole in the skull behind the eye)

T

tectonic - relating to 'plate tectonics,' known in common parlance as 'continental drift'

terrane - a small continental block or landmass

tetrapod - four-legged vertebrate (also includes descendants of four-legged vertebrates, such as snakes and whales)

trace fossil - a track, trail, or burrow left by an animal, e.g. a footprint; not a body fossil like a bone, shell, or leaf

trophic - relating to the food chain

U

ungulate - hoofed mammal

V

vagile - able to move around from place to place

vascular plants - plants possessing a water circulatory system; includes all "higher" plants; synonymous with tracheophytes

vertebrate - animal with a backbone

X

xeric - dry-adapted (usually referring to plants)

North America Through Time:

A variety of software and techniques was used to create the scenes at the beginnings of the sections and chapters, including 3D computer modeling and digital photography of models and real plants and animals. All of the backgrounds (landscapes) were created in Bryce — all the mountains, lakes, plains, rock formations, trees, ground covers, skies, and underwater landscapes are entirely computer-generated. Each separate object can be manipulated (sized, skewed, rotated, translated) and skins with color and texture added in the program until a satisfactory effect is achieved. The landscapes were then exported as Photoshop files and taken into Photoshop for the addition of animals, some plants, and finishing touches. The frames were styled usually in Typestyler but occasionally modified from Corel Gallery clip art in Photoshop. All individual components were then imported into Photoshop for compositing. Below are specific notes on each scene.

Hadean: scene done entirely in Bryce.

Archean: scene done entirely in Bryce.

Precambrian: background and stromatolites done in Bryce; ash cloud added from photograph.

Paleozoic: background done in Bryce; dragonfly from photograph of real dragonfly; *Eusthenopteron* from photo of model on display at University of Michigan Exhibit Museum of Natural History, Ann Arbor. Frame modified from Corel Gallery.

Cambrian: background done in Bryce; large *Anomalocaris*, *Canadaspis*, brown *Marrella*, and trilobites from photos of models on display at Royal Tyrrell Museum of Palaeontology, Drumheller, Alberta; small *Anomalocaris* from photo of Smithsonian travelling exhibit on the Burgess Shale when it was at the Sternberg Museum of Natural History, Hays, Kansas; school of orange *Marrella* and pink sponges from photo of models at University of Michigan Exhibit Museum of Natural History, Ann Arbor.

Ordovician: background, horn corals, and nautiloid shells done in Bryce; nautiloid cephalopod heads and seaweeds from photo of Ordovician diorama at University of Michigan Exhibit Museum of Natural History, Ann Arbor; large *Isotelus* trilobite at front center from photo (by Steve Brusatte) of model at National Museum of Natural History (Smithsonian); brown trilobite at far right from photo of real fossil.

Silurian: background done in Bryce; nautiloids from photos of models at Denver Museum of Nature & Science, Denver, Colorado; purple eurypterids from photo of model at University of Michigan Exhibit Museum of Natural History, Ann Arbor; orange eurypterid from photo at North American Museum of Ancient Life, Lehi, Utah; crinoids and horn corals from models at North American Museum of Ancient Life and Royal Tyrrell Museum of Palaeontology, Drumheller, Alberta; brachiopods from photo of real fossils.

Devonian: background done in Bryce; *Dunkleosteus* chasing school of *Osteolepis* from photo of diorama at North American Museum of Ancient Life, Lehi, Utah; *Pterolepis* (fairly large fish at center right) and *Pteraspis* (fish at upper left) from photos of models at American Museum of Natural History, New York; *Coccosteus* (fish at middle left) and *Pterichthyodes* (large fish at center bottom), horn corals, and crinoids from photos of models at University of Michigan Exhibit Museum of Natural History, Ann Arbor; trilobite and additional crinoids from photos of models at Denver Museum of Nature & Science, Denver, Colorado.

Mississippian: background done in Bryce; large coelacanth and large ammonite from photos of models at University of Michigan Exhibit Museum of Natural History, Ann Arbor; small coelacanth, small ammonites, crinoids, bryozoans, sharks, and brachiopods from photos of models at St. Louis Science Center, St. Louis, Missouri.

Pennsylvanian: scene done entirely in Bryce; millipede added from photo of model at North American Museum of Ancient Life, Lehi Utah; *Eryops* from photo of model on display at University of Michigan Exhibit Museum of Natural History, Ann Arbor. There are over 100 computer-generated prehistoric trees in this scene — *Calamites*, *Lepidodendron*, *Sigillaria*, *Cordaites*, and others all created using the tree lab.

Permian: background done in Bryce; *Dimetrodon* from photo of model by Allen A. Debus, author's collection; *Diplocaulus* from photo of model on display at Denver Museum of Nature & Science, Denver, Colorado.

Mesozoic: background done in Bryce; *Xiphactinus* fish and *Tylosaurus* mosasaur from photos of models on display at Sternberg Museum of Natural History, Hays, Kansas; ammonite from photo of model on display at Denver Museum of Nature & Science, Denver, Colorado; checkerboard border done in Photoshop.

Triassic: background done in Bryce; *Shonisaurus* ichthyosaur from photo of model on display at Las Vegas Museum of Natural History, Las Vegas, Nevada; ammonite from photo of model on display at North American Museum of Ancient Life, Lehi, Utah.

Jurassic: background and *Ginkgo* trees done in Bryce; cycads from photos of real cycads at Loran Whitelock's Cycad Gardens, Los Angeles, California; *Allosaurus* pack attacking *Diplodocus*, and *Archaeopteryx*, from photos of models on display at Burpee Museum of Natural History, Rockford, Illinois; *Rhamphorhynchus* from photo of model on display at North American Museum of Ancient Life, Lehi, Utah.

Cretaceous: background done in Bryce; pliosaur, hadrosaur, and pterosaurs from photos of models on display at Sternberg Museum of Natural History, Hays, Kansas; *T. rex* from photo of model on display at University of Michigan Exhibit Museum of Natural History, Ann Arbor.

Cenozoic: background done in Bryce; primates from photo of models on display at Denver Museum of Nature & Science, Denver, Colorado; condylarths from photo of models on display at University of Michigan Exhibit Museum of Natural History, Ann Arbor; frame modified from Corel Gallery.

Tertiary: background done in Bryce; uintatheres *Uintatherium* from photo of model on display at University of Michigan Exhibit Museum of Natural History, Ann Arbor; multituberculates are a chimaera of two pet rats we used to have, named Maple Sugar and Copper; frame modified from Corel Gallery.

Paleocene: background done in Bryce; pantodont is chimaera of photos of a capybara, a Dall sheep, and a kangaroo rat; multituberculate is modified from a photo of another of our pet rats, Oreo.

Eocene: background done in Bryce (corals created using tree lab); *Zygorhiza* whale from photo of model on display at Las Vegas Museum of Natural History, Las Vegas, Nevada; lobster is photo of my previous Everglades blue crayfish Yoda; scallops are photos of real fossils; fish and sea turtle are from photos taken at the Albuquerque Aquarium.

Oligocene: background done in Bryce; oreodonts, rhinos, horses, and camels are from photos of models on display at University of Michigan Exhibit Museum of Natural History, Ann Arbor.

Miocene: background done in Bryce; entelodont and gazelle-camels from photos of models on display at Denver Museum of Nature & Science, Denver, Colorado; horses, llama, and protoceratid are from photos of models on display at University of Michigan Exhibit Museum of Natural History, Ann Arbor.

Pliocene: background done in Bryce; glyptodont and mastodons from photos of models on display at University of Michigan Exhibit Museum of Natural History, Ann Arbor; horse from photo of model on display at Sternberg Museum of Natural History, Hays, Kansas; vulture from Adobe stock photos; pronghorns from photo of real animals (photo by Teddie Mobley).

Quaternary: background done in Bryce; mastodons from photo of models on display at University of Michigan Exhibit Museum of Natural History, Ann Arbor.

Pleistocene: background done in Bryce; large black mammoth at front left from photo of model outside Utah Field House of Natural History, Vernal, Utah; reddish mammoth at front right from photo of model on display at University of Michigan Exhibit Museum of Natural History, Ann Arbor; mammoths in background are from photos of a plastic set I have, given a Gaussian blur in Photoshop; caribou and fireweed are from photos I took in Alaska.

Holocene: background done in Bryce; eagle from Adobe stock photos; gaillardias, coyotes, bison, and mule deer from photos of real flowers and animals (photos by Teddie Mobley).

I am greatly indebted to the talented sculptors whose work is represented in these museums. Unfortunately, I cannot acknowledge them by name, as they were not identified by the museums displaying their work. Without their artistry, I would not have been able to create these scenes.

—LMC

Sources

Ch. 1: Precambrian

Alden, Andrew, 2000. Banded Iron Formation, http://geology.about.com/library/bl/images/blbif.htm

Anderson, Kristina, 2006. Ontario's Fossil Story, http://www.science.uwaterloo.ca/earth/waton/s0106.html

Anonymous, 2001. Banded Iron Formations, http://www.bookrags.com/sciences/earthscience/banded-iron-formations-woes-01.html

—, 2006. Iron oxide, http://en.wikipedia.org/wiki/Iron_oxide

Barton, Kate E. et al., 2003. The North America Tapestry of Time and Terrain, http://www.tapestry.usgs.gov

Behrendt, Marc, 2003. Feature Fossil: stromatolite. Fossil News 9 (12): 3-4.

Briggs, Derek E. G., and Richard Fortey, 2005. Wonderful Strife: Systematics, Stem Groups, and the Phylogenetic Signal of the Cambrian Radiation. in Macroevolution: Diversity, Disparity, Contingency, E. S. Vrba and N. Eldredge (eds), Supplement to Paleobiology v. 31 (2).

Collins, Allen G. and Ben M. Waggoner, 1994a. Introduction to the Vendian Period, http://www.ucmp.berkeley.edu/vendian/vendian.html

—, 1994b. Learning about Vendian animals, http://www.ucmp.berkeley.edu/vendian/critters.html

—, 1994c. Localities of the Vendian: Ediacara Hills, Australia, http://www.ucmp.berkeley.edu/vendian/ediacara.html

Cowen, Richard, 2005. History of Life. Blackwell Publishing, Oxford, UK.

Delwiche, R., 2006. Photosynthetic Life in the Fossil Record, http://www.life.umd.edu/labs/delwiche/PSlife/lectures/Fossils.html

Donnadieu, Y. et al., 2004. Continental Breakup and "Snowball Earth," http://scienceweek.com/2004/sa040730-5.htm

Encyclopedia Britannica Online, 2006. Origin of banded iron formation, http://www.britannica.com/eb/article-80298

Hagadorn, James W. and Ben M. Waggoner, 2000. Ediacaran Fossils from the Southwestern Great Basin, United States. Journal of Paleontology 74 (2): pp. 349-359.

Hibbard, James et. al, 2006. Significance of a new Ediacaran fossil find in the Carolina Terrane of North Carolina, http://gsa.confex.com/gsa/2006NE/finalprogram/abstract_99453.htm

Hofmann, Hans J., 1998. Synopsis of Precambrian Fossil Occurrences in North America (and references therein). Chapter 4 in Geology of the Precambrian Superior and Grenville provinces and Precambrian fossils in North America, S. B. Lucas and M. R. St-Onge (coordinators), Geology of Canada no. 7, pp. 271-376, Geological Survey of Canada; also published under the same title by Geological Survey of America, The Geology of North America, v. C-1, pp. 271-376

—, 2004. Precambrian fossils in Québec. pp. 109-113 in G. Prichonnet and M. A. Bouchard (eds), Actes du premier colloque du patrimonine géologique du Québec, Géologie Québec MB 2004-05.

Hoover, Aaron, 2004. Supercontinent's Breakup Plunged Ancient Earth into Big Chill, University of Florida News, http://www.napa.ufl.edu/2004news/snowballearth.htm

Ingham, Richard, 2004. How the Earth became a Snowball, News in Science, http://www.abc.net.au/science/news/stories/s1068501.htm

Irion, Robert, 2004. Humble Pocket Gophers Shed Light on the Genetic Fortitude of Cheetahs, http://lynx.uio.no/jon/lynx/cheetahg.htm

Kazlev, M. Alan, 2002. The Precambrian, http://www.palaeos.com/Timescale/Precambrian.htm

King, Philip B., 1977. The Evolution of North America. Princeton University Press, Princeton, NJ.

Klarreich, Erica, 2005. Life on the Scales. Science News v. 167, pp. 106-108, February 12.

Kohlbecker, Matthew V., 2006. Banded Iron Formations (BIFs): characteristics, modes of occurrence, and geological significance. Denison Journal of the Geosciences, http://www.denison.edu/geology/geojournal/kohlbecker.html

Kudryavtsev, Anatoliy B. et. al, 2001. In situ laser-Raman imagery of Precambrian microscopic fossils, Proceedings of the National Academy of Sciences 98 (3): pp. 823-826.

Lipps, Jere H., 2002. The Radiation of the First Animals, http://www.accessexcellence.org/BF/bf02/lipps/

Margulis, Lynn, Michael F. Dolan, and Jessica H. Whiteside, 2005. "Imperfections and Oddities" in the Origin of the Nucleus. in Macroevolution: Diversity, Disparity, Contingency, E. S. Vrba and N. Eldredge (eds), Supplement to Paleobiology v. 31 (2).

McMenamin, Mark A. S., 1998. The Garden of Ediacara. Columbia University Press, New York.

Metz, Karen, 2001. Late Proterozoic Plate Tectonic and Paleo-Environmental Reconstructions: Implications for the Snowball Earth Hypothesis, http://gsa.confex.com/gsa/2001ESP/finalprogram/abstract_7961.htm

Michigan State University, 2006. The Precambrian Era, http://www.geo.msu.edu/geo333/Precambrian.html

Misra, S. B., 1969. Pioneering Work on the Mistaken Point Fauna. Geological Society of America Bulletin v. 80, pp. 2133-2140.

Morrison, Reg, 2002. Australia: Land Beyond Time. Comstock Publishing, Ithaca, NY.

Murphy, J. B. and R. D. Nance, 2004. On the Assembly of Supercontinents, http://scienceweek.com/2004/sa040730-5.htm

National Park Service, 2006. Glacier National Park, http://www.nps.gov/glac/pphtml/subnaturalfeatures16.html

Newfoundland & Labrador Parks Division, 2006. The Fortune Head Ecological Reserve, http://www.k12.nf.ca/lakeacademy/fossil2.htm

O'Brien, S. J., A. F. King, and H. J. Hofmann, 2006. Lithostratigraphic and biostratigraphic studies on the eastern Bonavista Peninsula: an update. Current Research, Newfoundland and Labrador Dept. of Natural Resources Geological Survey, Report 06-1, pp. 257-263.

Ogg, James, 2003. Overview of Global Boundary Stratotype Sections and Points (GSSP's). http://www.stratigraphy.org/gssp.htm

Peterson, Kevin J., Mark A. McPeek, and David A. D. Evans, 2005. Tempo and Mode of Early Animal Evolution: Inferences from Rocks, Hox, and Molecular Clocks. in Macroevolution: Diversity, Disparity, Contingency, E. S. Vrba and N. Eldredge (eds), Supplement to Paleobiology v. 31 (2).

Rohde, Robert A., 2004. GeoWhen Database, http://www.stratigraphy.org/geowhen/index.html

Scotese, Christopher, 2004. Paleomap Project, http://www.scotese.com/precambr.htm and http://www.scotese.com/info.htm

Speer, Brian R., 1997. Introduction to the Archean, http://www.ucmp.berkeley.edu/precambrian/archaean.html

Strick, M., 2006. Banded Iron Formation, http://jersey.uoregon.edu/~mstrick/RogueComCollege/RCC_Lectures/Banded_Iron.html

Torsvik, Trond H., 2004. On the Supercontinent Rodinia, http://scienceweek.com/2004/sa040730-5.htm

Waggoner, Ben M., 1996. Hadean time, http://www.ucmp.berkeley.edu/precambrian/hadean.html

—, 2001. Localities of the Vendian: Mistaken Point, Newfoundland, http://www.ucmp.berkeley.edu/vendian/mistaken.html

Waggoner, Ben M. and Brian R. Speer, 1997a. Introduction to the Proterozoic Era, http://www.ucmp.berkeley.edu/precambrian/proterozoic.html

—, 1997b. Proterozoic Era: Life, http://www.ucmp.berkeley.edu/precambrian/proterolife.html

Ch. 2: Cambrian

Anonymous, 2004. The Cambrian Mass Extinction, http://hannover.park.org/Canada/Museum/extinction/cammass...

—, 2004. The Cambrian Period, http://www.palaeos.com/Paleozoic/Cambrian/Cambrian.htm

—, 2007. Fossils in Death Valley National Park, http://members.aol.com/Waucoba5/dv/dvfossils.htm#virtualfieldtrips

Barton, Kate E. et al., 2003. The North America Tapestry of Time and Terrain, http://www.tapestry.usgs.gov

Behrendt, Marc, 2000. Feature Fossil: Hyolithes. Fossil News 6 (10): 3-4.

—, 2002a. Feature Fossil: Brooksella alternata. Fossil News 8 (3): 3.

—, 2002b. Feature Fossil: Selenocoryphe platyura. Fossil News 8 (4): 3.

—, 2002c. Feature Fossil: Modocia laevanucha. Fossil News 8 (12): 3-4.

—, 2003a. The Weeks Formation, House Range, Utah. Fossil News 9 (9): 15-19.

—, 2003b. Feature Fossil: Glyptagnostus reticulatus. Fossil News 9 (10): 3.

—, 2003c. Feature Fossils: *Climactichnites* and *Diplichnites*. Fossil News 9 (11): 3-4.

Budd, Graham E., 2003. The Cambrian Fossil Record and the Origin of Phyla. Integrative and Comparative Biology 43 (1): pp. 157-165.

Burgess Shale Geoscience Foundation website, http://www.burgess-shale.bc.ca

Ciampaglio, Charles N., 2006. Reinterpretation of *Brooksella* from the Conasauga Formation (Cambrian) of Georgia and Alabama, USA. Geological Society of America Abstracts with Programs 38 (3): p. 4 (paper no. 2-1).

Clos, Lynne M., 2003. Field Adventures in Paleontology. Fossil News, Boulder, CO.

Clowes, Christopher, 2004a. The Cambrian Period, http://www.peripatus.gen.nz/paleontology/Cambrian.html

—, 2004b. The Cambrian "Explosion," http://www.peripatus.gen.nz/paleontology/CamExp.html

—, 2007. Sirius Passet Lagerstätte, http://www.peripatus.gen.nz/Paleontology/lagSirPas.html

Collins, Allen G., 1994a. Metazoa: Fossil Record, http://www.ucmp.berkeley.edu/phyla/metazoafr.html

—, 1994b. Archaeocyatha, http://www.ucmp.berkeley.edu/porifera/archaeo.html

Garcia-Bellido Capdevila, D. and S. Conway Morris, 1999. New Fossil Worms from the Lower Cambrian of the Kinzers Formation, Pennsylvania, with Some Comments on Burgess Shale-Type Preservation. Journal of Paleontology 73 (3): pp. 394-402.

Geyer, G., 1997. The First Shelly Faunas, http://wwwalt.uni-wuerzburg.de/palaeontologie/Stuff/casu8.htm

Gunther, Lloyd F. and Val G. Gunther, 1981. Some Middle Cambrian Fossils of Utah. Brigham Young University Geology Studies 28 (1), Provo, UT.

Hagadorn, James W., Christopher M. Fedo, and Ben M. Waggoner, 2000. Early Cambrian Ediacaran-Type Fossils from California. Journal of Paleontology 74 (4): pp. 731-740.

Hunt, C. B. and Mabey, D. R., 1966. General Geology of Death Valley, California. U.S.G.S. Professional Paper no. 494, U.S. Geological Survey.

Idaho State University, 2007. Fossils in Idaho, http://imnh.isu.edu/digitalatlas/geo/fossils/fossils.htm

Kimball, J., 2004. The Geologic and Evolutionary Record, http://users.rcn.com/jkimball.ma.ultranet/BiologyPages/G/Geo...

Lieberman, Bruce S., 2003. A New Soft-Bodied Fauna: the Pioche Formation of Nevada. Journal of Paleontology 77 (4).

MacRae, Andrew, 2007. Burgess Shale Fossils, http://www.geo.ucalgary.ca/~macrae/Burgess_Shale/

Milius, Susan, 2007. Primate's Progress. Science News v. 171, April 14.

Mount, Jack D., 1980. Characteristics of Early Cambrian Faunas from Eastern San Bernardino County, California. pp. 19-29 in Paleontological Tour of the Mojave Desert, California-Nevada, Southern California Paleontological Society Special Publication no. 2.

Ogg, James, 2003. Overview of Global Boundary Stratotype Sections and Points (GSSP's) http://www.stratigraphy.org/gssp.htm

Poirier, Ben, 2001. The Search for Ruin Wash Trilobites. Fossil News 7 (8): 6-8.

Scotese, Christopher, 2000a. Early Cambrian Climate, http://www.scotese.com/ecambcli.htm

—, 2000b. Late & Middle Cambrian Climate, http://www.scotese.com/mlcambcl.htm

Scott, Erin et al., 2000. Localities of the Cambrian: the Burgess Shale, www.ucmp.berkeley.edu/cambrian/burgess.html

Stinchcomb, Bruce L., 1997. Missouri Fossils, Rocks & Minerals, November.

Valentine, James W., 2004. On the Origin of Phyla. University of Chicago Press, Chicago, IL.

Waggoner, Ben, 2000a. Localities of the Cambrian: the Marble Mountains, http://www.ucmp.berkeley.edu/cambrian/marblemts.html

—, 2000b. Localities of the Cambrian: the House Range, http://www.ucmp.berkeley.edu/cambrian/house.html

—, 2001. Life During the Cambrian Period, http://www.ucmp.berkeley.edu/cambrian/camblife.html

Westrop, Stephen R. and Ed Landing, 2000. Lower Cambrian (Branchian) Trilobites and Biostratigraphy of the Hanford Brook Formation, Southern New Brunswick. Journal of Paleontology 74 (5): 858-878.

Whiteley, Thomas E., Gerald J. Kloc, and Carlton E. Brett, 2002. Trilobites of New York. Comstock Publishing, Ithaca, NY.

Whittington, Harry B., 1985. <u>The Burgess Shale</u>. Yale University Press, New Haven, CT.

Xian-Guang, Hou et al., 2004. <u>The Cambrian Fossils of Chengjiang, China</u>. Blackwell Publishing, Oxford, UK.

Ch. 3: Ordovician

Avildsen, Christina, Jennifer Bie, Chirag Patel, and Brie Sarvis, 1998. The Ordovician, http://www.ucmp.berkeley.edu/ordovician/ordovician.html

Anonymous, 2002. Fossil Bryozoan from Southern Ontario, Canada, http://www.turnstone.ca/bryozoan.htm

—, 2004a. The Ordovician Mass Extinction, http://hannover.park.org/Canada/Museum/extinction/ordmass.html

—, 2004b. Speculated Causes of the Ordovician Extinction, http://hannover.park.org/Canada/Museum/extinction/ordcause.html

—, 2004c. The Ordovician Period, http://www.palaeos.com/Paleozoic/Ordovician/Ordovician.htm

—, 2007a. Fossil Mountain, http://ammonoid.com/fossilmt.htm

—, 2007b. Fossil Formation, http://www.nashvillefossils.com/exercises/formation/formation.html

—, 2007c. Ordovician Brachiopods...Indiana & Ohio, http://www.geocities.com/atrypa/brach6.html

—, 2007d. Ordovician Fossils in the Toquima Range, Nevada, http://members.aol.com/Waucoba7/tr/toquima.html

—, 2007e. Field Trip to the Great Beatty Mudmound/Bioherm, http://members.aol.com/Waucoba7/beatty/beattyfossils.html

Barnes, Christopher R., 2004. Ordovician Oceans and Climate. Ch. 7 in B. D. Webby et al. (eds), <u>The Great Ordovician Biodiversification Event</u>. Columbia University Press, New York.

Barton, Kate E. et al., 2003. The North America Tapestry of Time and Terrain, http://www.tapestry.usgs.gov

Behrendt, Marc, 2007. Trilobites of the Cincinnati Region. Fossil News 13 (5): 2-10.

Brusatte, Stephen, 2002. <u>Stately Fossils</u>. Fossil News, Boulder, CO.

Carrera, Marcela G. and J. Keith Rigby, 2004. Sponges. Ch. 12 in B. D. Webby et al. (eds), <u>The Great Ordovician Biodiversification Event</u>. Columbia University Press, New York.

Chaper, Joseph, 2007. Geology of Missouri—Ordovician, http://members.sockets.net/~joschaper/ordo.html

Clowes, Christopher, 2004. Ordovician Period, http://www.peripatus.gen.nz/paleontology/Ordovician.html

Cocks, L. Robin and Trond H. Torsvik, 2004. Major Terranes of the Ordovician. Ch. 5 in B. D. Webby et al. (eds), <u>The Great Ordovician Biodiversification Event</u>. Columbia University Press, New York.

Cooper, Roger A. et al, 2004. Graptolites: Patterns of Diversity Across Paleolatitudes. Ch. 27 in B. D. Webby et al. (eds), <u>The Great Ordovician Biodiversification Event</u>. Columbia University Press, New York.

Cowen, Richard, 2005. <u>History of Life</u>. Blackwell Publishing, Oxford, UK.

Edwards, D., 2001. Early Land Plants, pp. 63-66 in: <u>Palaeobiology II</u>, Briggs, D. and Crowther, P. (eds), Blackwell Science, Oxford, UK.

Frey, Robert C. et al., 2004. Nautiloid Cephalopods. Ch. 21 in B. D. Webby et al. (eds), <u>The Great Ordovician Biodiversification Event</u>. Columbia University Press, New York.

Gore, Pamela J. W., 1999. The Ordovician Period. http://www.de/peachnet.edu/~pgore/geology/geo102/ordo.htm

Guiliermo, L. Albanesi and Stig M. Bergstrom, 2004. Conodonts: Lower to Middle Ordovician Record. Ch. 29 in B. D. Webby et al. (eds), <u>The Great Ordovician Biodiversification Event</u>. Columbia University Press, New York.

Hansen, Michael C., 1997. The Geology of Ohio—the Ordovician. Ohio Geology, fall issue.

Hermann, Achim D., 2003. Linking Early Late Ordovician paleobiogeographic data with climate-ocean models. http://gsa.confex.com/gsa/2003AM/finalprogram/abstract_59855.htm

—, and Mark E. Patzkowsky, 2003. Ocean and climate model results and the Late Ordovician mass extinction. http://www.unt.edu.ar/fcsnat/INSUGEO/geologia%20_17/88.htm

Hess, Hans et al., 1999. <u>Fossil Crinoids</u>, Cambridge University Press, Cambridge, UK.

Idaho State University, 2007. Fossils in Idaho, http://imnh.isu.edu/digitalatlas/geo/fossils/fossils.htm

Karim, Talia and Stephen R. Westrop, 2002. Taphonomy and Paleoecology of Ordovician Trilobite Clusters, Bromide Formation, South-central Oklahoma. Palaios 17(4): 394-402.

Kohl, Martin, 2007. Welcome to the Gray Site, http://www.state.tn.us/environment/tdg/gray/

Long, John A., 1995. The Rise of Fishes. The Johns Hopkins University Press, Baltimore, MD.

Mángano, M. Gabriela and Mary L. Droser, 2004. The Ichnologic Record of the Ordovician Radiation. Ch. 34 in B. D. Webby et al. (eds), The Great Ordovician Biodiversification Event. Columbia University Press, New York.

McKinstry, Lea H. et al., 2001. Enigmatic Fossils from the Martinsburg Formation (Upper Ordovician), Northeastern Tennessee, http://gsa.confex.com/gsa/2001SE/finalprogram/abstract_4546.htm

Nova Scotia Museum, Fossils of Nova Scotia. http://museum.gov.ns.ca/fossils/geol/ordo.htm

Ogg, James, 2003. Overview of Global Boundary Stratotype Sections and Points (GSSP's). http://www.stratigraphy.org/gssp.htm

Pestana, Harold R., 1960. Fossils from the Johnson Spring Formation, Middle Ordovician, Independence Quadrangle, California. Journal of Paleontology 34 (5): 862-873.

Poling, Jeff, 1996. Maps of Ancient Earth. http://www.dinosauria.com/dml/maps.htm

Retallack, Gregory J. and Carolyn R. Feakes, 1987. Trace Fossil Evidence for Late Ordovician Animals on Land. Science 237 (4784): 61-63.

Schrantz, Richard, 2000. Photographs of Fossils Found on KPS Field Trips, http://www.uky.edu/OtherOrgs/KPS/pages/fossilphoto.html

Schumacher, Gregory A. and William I. Ausich, 1983. New Upper Ordovician Echinoderm Site: Bull Fork Formation, Caesar Creek Reservoir (Warren County, Ohio). Ohio Journal of Science 83 (1): 60-64.

Scotese, Christopher, 2000. Early Ordovician Climate. http://www.scotese.com/eordclim.htm

Shaw, Jeremy, 2007. Ordovician Trentonian Fauna from the Galena and Maquoketa of West De Pere and Green Bay, Wisconsin, http://www.fox.uwc.edu/fossils/wisc/

Stanley, D. Christopher and Ron K. Pickerill, 1998. Review of Systematic Ichnology of the Late Ordovician Georgian Bay Formation of Southern Ontario, Eastern Canada, http://www.envs.emory.edu/ichnology/IN99-MACNAU~1.HTM

Stanley, Thomas M. and Rodney M. Feldmann, 1998. Significance of Nearshore Trace-Fossil Assemblages of the Cambro-Ordovician Deadwood Formation and Aladdin Sandstone, South Dakota. Annals of Carnegie Museum 67 (1): 1-51.

Steemans, Philippe and Charles H. Wellman, 2004. Miospores and the Emergence of Land Plants. Ch. 33 in B. D. Webby et al. (eds), The Great Ordovician Biodiversification Event. Columbia University Press, New York.

Stinchcomb, Bruce L., 19977. Missouri Fossils. Rocks & Minerals 7(6), November.

Stitt, James H., 2000. Additional Information on Lowest Ordovician Trilobites from the Uppermost Deadwood Formation, Black Hills and Bear Lodge Mountains, South Dakota and Wyoming. Journal of Paleontology, March.

Turner, Susan, Alain Blieck, and Godfrey S. Nowlan, 2004. Vertebrates (Agnathans and Gnathostomes). Ch. 30 in B. D. Webby et al. (eds), The Great Ordovician Biodiversification Event. Columbia University Press, New York.

Webby, Barry D., 2004. Introduction (to the Ordovician). Ch. 1 in B. D. Webby et al. (eds), The Great Ordovician Biodiversification Event. Columbia University Press, New York.

—, Roger A. Cooper, Stig M. Bergstrom and Florentin Paris, 2004. Stratigraphic Framework and Time Slices. Ch. 2 in B. D. Webby et al. (eds), The Great Ordovician Biodiversification Event. Columbia University Press, New York.

—, and Robert J. Elias et al., 2004. Corals. Ch. 15 in B. D. Webby et al. (eds), The Great Ordovician Biodiversification Event. Columbia University Press, New York.

Westphal, Klaus W., 1974. New Fossils from the Middle Ordovician Platteville Formation of Southwest Wisconsin. Journal of Paleontology 48 (1): 78-83.

Whiteley, Thomas E., Gerald J. Kloc, and Carlton E. Brett, 2002. Trilobites of New York. Comstock Publishing, Ithaca, NY.

Ch. 4: Silurian

Anonymous, 2004a. Silurian Geology, http://www.silurian.com/geology/

—, 2004b. Oklahoma Geology and Paleontology, http://www.freewebz.com/oklahomarocks/silurian.htm

—, 2004c. Silurian Period, http://www.uky.edu/KGS/coal/webfossl/pages/silurian.htm

—, 2007. Field Trips to Napoleon, Indiana and Owingsville, Kentucky, http://www.uky.edu/OtherOrgs/KPS/pages/napoleon.html

Barton, Kate E. et al., 2003. The North America Tapestry of Time and Terrain, http://www.tapestry.usgs.gov

Behrendt, Marc, 2000. Feature fossil: *Caryocrinites ornatus*. Fossil News 6 (2): p. 3.

—, 2001a. Feature fossil: *Eucalyptocrinites tuberculatus*. Fossil News 7 (3): p. 3.

—, 2001b. Feature fossil: bryozoan. Fossil News 7 (6): p. 3.

—, 2003a. Feature fossils: *Dalmanites limulurus* and *Calymene niagarensis*. Fossil News 9 (2): p. 3.

—, 2003b. Feature fossil: *Eurypterus remipes*. Fossil News 9 (4): p. 3.

Chiang, Kam K., 1971. Silurian Pentameran Brachiopods of the Fossil Hill Formation, Ontario. Journal of Paleontology 45(5): 849-861.

Ciurca, Samuel J., 2000. Prehistoric Pittsford: the Silurian Eurypterid Fauna. http://www.eurypterid.net

Clowes, Christopher, 2004. Silurian Period, http://www.peripatus.gen.nz/paleontology/Silurian.html

De Freitas, T. A. et al., c. 2000. Silurian System of the Canadian Arctic Archipelago. http://www.cspg.org/Vol47No2deFreitas.pdf

DeWindt, J. Thomas, 1973. Occurrence of *Rusophycus* in the Poxono Island Formation (Upper Silurian) of Eastern Pennsylvania. Journal of Paleontology 47(5): 999-1000.

DiMichele, William A. and Robert W. Hook (rapporteurs), 1992. Paleozoic Terrestrial Ecosystems. Ch. 5 in Anna K. Behrensmeyer et al. (eds), Terrestrial Ecosystems Through Time. University of Chicago Press, Chicago, IL.

Edwards, D., 2001. Early Land Plants, pp. 63-66 in: Palaeobiology II, Briggs, D. and Crowther, P. (eds), Blackwell Science, Oxford, UK.

Edwards, Jonathan, 1981. A Brief Description of the Geology of Maryland. http://www.mgs.md.gov/esic/brochures/mdgeology.html

Falls of the Ohio State Park website, http://www.fallsoftheohio.org/collecting.html

Fischer, Dan, Tammy Liu, Emily Yip, and Korsen Yu, 1998a. The Silurian, http://www.ucmp.berkeley.edu/silurian/silurian.html

—, 1998b. Life of the Silurian, http://www.ucmp.berkeley.edu/silurian/silulife.html

Hagström, Jonas, 2003. Biostratigraphy Based on the Spores of Early Land Plants. http://www.nrm.se/pb/research/jh_silurianeng.html

Hanlon, Julie, 2001. The Geological History of Eastern New York State between the Late Silurian and Middle Devonian Periods, http://www.sas.upenn.edu/earth/geol205/geol_hist.html

Idaho State University, 2007. Fossils in Idaho, http://imnh.isu.edu/digitalatlas/geo/fossils/fossils.htm

International Commission on Stratigraphy, 2004a. GSSP for the Ordovician-Silurian Boundary, http://www.stratigraphy.org/ordsil.htm

—, 2004b. GSSP for the Silurian-Devonian Boundary, http://www.stratigraphy.org/sildev.htm

Joyce, Stacey, 2000. Origin and Composition of Silurian Stromatolites, Glacier Bay National Park, Alaska. Colgate University Journal of the Sciences, pp. 85-114.

Kerp, Hans, 2000. A History of Paleozoic Forests. http://www.biologie.uni-hamburg.de/b-online/kerp/ewald.html

Ljuboja, Zarko, 2007. The Fauna and Flora of the Silurian Fiddlers Green Formation and Association with *Eurypterus remipes* of Herkimer County, http://www.langsfossils.com/pages/edu-2.htm

Long, John A., 1995. The Rise of Fishes. The Johns Hopkins University Press, Baltimore, MD.

—, 2001. Rise of Fishes, pp. 52-57 in: Palaeobiology II, Briggs, D. and Crowther, P. (eds), Blackwell Science, Oxford, UK.

Maryland Geological Survey, 2004. Maryland Fossils, http://www.mgs.md.gov/esic/brochures/fossils/devofos.html

Massare, J., 2004a. Rochester Shale, http://www.weather.brockport.edu/~jmassare/silurian/rochester.htm

—, 2004b. Lockport Formation, http://www.weather.brockport.edu/~jmassare/silurian/lockport.htm

Milwaukee Public Museum, 2004a. Silurian Reefs in the Great Lakes Region, http://www.mpm.edu/reef/great-lakes-reefs.html

—, 2004b. Silurian Reefs in Southeastern Wisconsin and Northeastern Illinois, http://www.mpm.edu/reef/wi-il-reefs.html

Ogg, James, 2003. Overview of Global Boundary Stratotype Sections and Points (GSSP's). http://www.stratigraphy.org/gssp.htm

Peel, J.S., 1977. Systematics and Paleoecology of the Silurian Gastropods of the Arisaig Group, Nova Scotia. Det. Kong. Danske Viden. Selskab 21(2): 80.

Phelan, Sean P., 1998. Silurian Geology of Western New York State. http://www.weather.brockport.edu/~jmassare/silurian/silurian.htm

Sam Noble Oklahoma Museum of Natural History, 2004. Silurian, http://www.snomnh.ou.edu/COLLECTIONS&RESEARCH/sub/common_fossils_of_ok/Silurian1.htm

Scotese, Christopher, 2000a. Silurian Climate. http://www.scotese.com/silclim.htm

—, 2000b. Continents Begin to Collide as Paleozoic Oceans Close. http://www.scotese.com/newpage2.htm

Shrake, Douglas L., 2000. Fossil Collecting in Ohio. GeoFacts no. 17, Ohio Department of Natural Resources, Division of Geological Survey.

Stemmerik, Lars et al., 1997. Palaeo-oil Field in a Silurian Carbonate Buildup, Wulff Land, North Greenland: Project 'Resources of the Sedimentary Basins of North and East Greenland.' Geology of Greenland Survey Bulletin 176: 24-28.

Stewart, Wilson N. and Rothwell, Gar W., 1993. Paleobotany and the Evolution of Plants, 2nd ed. Cambridge University Press, Cambridge, UK.

Tomescu, Julia and Kidder, David L., 2002. Ordovician and Silurian Cherts and their Relation to Paleoclimate, Paleoceanography, and Onshore-Offshore Evolution, http://gsa.confex.com/gsa/2002NC/finalprogram/abstract_29874.htm

University of California, Berkeley, 2004. The Silurian Period in Pennsylvania, http://www.ucmp.berkeley.edu/tapestry/time_and_space/states/PA_S.shtml

Whiteley, Thomas E., Kloc, Gerald J., and Brett, Carlton E., 2002. Trilobites of New York. Comstock Publishing Assoc. (a division of Cornell University Press), Ithaca, NY.

Wilde, Pat et al., 1991. Silurian Oceanic and Atmospheric Circulation and Chemistry, http://users.rcn.com/patwilde/sil91.html

Wozniak, Carl, 2004. The Silurian, http://www.carlwozniak.com/earth/Silurian.html

Ch. 5: Devonian

Anonymous, 2004a. The Devonian Mass Extinction, http://hannover.park.org/Canada/Museum/extinction/devmass.html

—, 2004b. Causes of the Devonian Extinction, http://hannover.park.org/Canada/Museum/extinction/devcause.html

—, 2004c. Devonian Age in Kentucky, http://www.uky.edu/KGS/coal/webfossil/pages/devonian.htm

—, 2004d. Devonian Paleogeography, Southwestern US, http://jan.ucc.nau.edu/~rcb7/devpaleo.html

—, 2007a. Fossils Found at the Ludlowville and Moscow Formations at the Lake Erie Cliffs on the Mouth of 18 Mile Creek in New York, http://www.fossilguy.com/sites/18mile/18_col.htm

—, 2007b. Oklahoma Trilobites, http://www.fossilmall.com/EDCOPE_Enterprises/trilobites/OK1.htm

—, 2007c. Field Trip to Parsons, http://www.nashvillefossils.com/fieldtrips/parsonsfieldtrip/parsons.html

Barton, Kate E. et al., 2003. The North America Tapestry of Time and Terrain, http://www.tapestry.usgs.gov

Behrendt, Marc, 2000a. Feature fossil: phyllocarid. Fossil News 6 (3): p. 3.

—, 200b. Feature fossil: Platyceras multispinosum. Fossil News 6 (4): p. 3.

—, 2001a. Feature fossil: Coccosteus cuspidata. Fossil News 7 (1): p. 3.

—, 2001b. Feature fossils: Phacops rana crassituberculata and Eldredgeops milleri. Fossil News 7 (2): pp. 3-4.

—, 2001c. Feature fossil: Devonian sea slab. Fossil News 7 (9): p. 3.

—, 2002. Feature fossil: Devonian shark tooth. Fossil News 8 (5): p. 3.

—, 2003. Feature fossil: Hyneria lindae. Fossil News 9 (9): pp. 3-4.

Boggs, Sam Jr., 1987 Principles of Sedimentology and Stratigraphy. Merrill Publishing Co., Columbus, OH.

Bond, David and Paul Wignall, 2003. Frasnian/Famennian Boundary Anoxia in the Great Basin, Western United States. http://gsa.confex.com/gsa/2003AM/finalprogram/abstract_59909.htm

Brusatte, Stephen, 2002. Stately Fossils: a Comprehensive Look at the State Fossils and Other Official Fossils. Fossil News, Boulder, CO.

Clos, Lynne M., 2003. Field Adventures in Paleontology. Fossil News, Boulder, CO.

Clowes, Christopher, 2004. Devonian Period, http://www.peripatus.gen.nz/paleontology/Devonian.html

Coates, M. I., 2001. Origin of Tetrapods. in Paleobiology II, Briggs, D. and Crowther, P. (eds), Blackwell Science, Oxford, UK.

Cowen, Richard, 2005. History of Life. Blackwell Publishing, Oxford, UK.

Daeschler, Edward B., 2003. Biogeographic Patterns of Late Devonian Vertebrates from North America. http://gsa.confex.com/gsa/2003AM/finalprogram/abstract_60394.htm

Day, Jed, 2003. Western Laurussian Record of Regional and Global Middle-Late Devonian Sea Level Changes and Bioevents: Alberta Rocky Mountains. http://gsa.confex.com/gsa/2003AM/finalprogram/abstract_60231.htm

DiMichele, William A. and Robert W. Hook (rapporteurs), 1992. Paleozoic Terrestrial Ecosystems, ch. 5 in Anna K. Behrensmeyer et al. (eds), Terrestrial Ecosystems through Time, University of Chicago Press, Chicago, IL.

Edwards, D., 2001. Early Land Plants. in Paleobiology II, Briggs, D. and Crowther, P. (eds), Blackwell Science, Oxford, UK.

Etter, Walter, 2002. Hunsrück Slate: Widespread Pyritization of a Devonian Fauna. in Bottjer, D. et al (eds), Exceptional Fossil Preservation. Columbia University Press, New York, NY.

Falls of the Ohio State Park website, http://www.fallsoftheohio.org

Glausiusz, Josie, 1999. The Old Fish of the Sea. Discover magazine, January 1999.

Hill, Richard E., 2007, pers. comm.

Itano, Wayne, 1998. Colorado Devonian Fossils. http://www.itano.net/fossils/chaffee/chaffee.htm

Jensen, Mari N., 1998. Modern Climate Has Roots in Early Devonian. Science News v. 159, 2/14/98.

Joachimski, M. M. and W. Buggisch, 2000. The Late Devonian Mass Extinction—Impact or Earth-Bound Event? http://www.lpi.usra.edu/meetings/impact2000/pdf/3072.pdf

Johnson, Kirk, and Richard Stucky, 1995. Prehistoric Journey: a History of Life on Earth. Roberts Rinehart Publishers, Boulder, CO.

Kerp, Hans, 2000. A History of Paleozoic Forests. http://www.biologie.uni-hamburg.de/b-online/kerp/ewald.html

Kesling, Robert V., 1982. Arkonaster, a New Multi-Armed Starfish from the Middle Devonian Arkona Shale of Ontario. Contributions from the Museum of Paleontology, University of Michigan 26(6): 83-115.

Long, John A., 1995. The Rise of Fishes. The Johns Hopkins University Press, Baltimore, MD.

Marshall, J. E. A., 2003. The Taghanic Event: a Late Devonian Aridity Crisis. http://gsa.confex.com/gsa/2003AM/finalprogram/abstract_57417.htm

McGhee, George R., Jr., 2003. Testing Late Devonian Extinction Hypotheses. http://gsa.confex.com/gsa/2003AM/finalprogram/abstract_57883.htm

Moore, Raymond C., Cecil G. Lalicker, and Alfred G. Fischer, 1952. Invertebrate Fossils (ch. 9: cephalopods). McGraw-Hill, New York.

Murphy, Dennis, 2002. Devonian Times. http://www.mdgekko.com/devonian

NOVA: Following Scientists' Pursuit of a Big, Ugly, and Very Ancient Fish Called the Coelacanth. PBS-TV, broadcast 4/6/04.

Nova Scotia Museum, 2004. Fossils of Nova Scotia: Devonian Period, http://museum.gov.ns.ca/fossils/geol/devo.htm

Ogg, Jim, 2003. Overview of Global Boundary Sections and Points (GSSP's). http://www.stratigraphy.org/gssp.htm

Perkins, Sid, 2004. Early Flight? Winged Insects Appear Surprisingly Ancient. Science News v. 165, 2/14/04.

—, 2007. Forest Primeval: the Oldest Known Trees Finally Gain a Crown. Science News, v. 171, 4/21/07, p. 243-44.

Racki, Grzegorz, 2003. Toward Understanding of Late Devonian Global Events: Few Answers, Many Questions, http://gsa.confex.com/gsa/2003AM/finalprogram/abstract_61972.htm

Ross, Kelley, 1997. The Great Devonian Controversy: the Shaping of Scientific Knowledge Among Gentlemanly Specialists, http://www.friesian.com/rudwick.htm

Scheckler, Stephen E. 2001. Afforestation—the First Forests. in Paleobiology II, Briggs, D. and Crowther, P. (eds), Blackwell Science, Oxford, UK.

—, 2003. Consequences of Rapid Expansion of Late Devonian Forests. http://gsa.confex.com/gsa/2003AM/finalprogram/abstract_65680.htm

Schieber, Juergen, 2003. Disturbed Beds in Lower Frasnian Black Shales of Tennessee and Kentucky: Their Possible Connection to Impacts. http://gsa.confex.com/gsa/2003AM/finalprogram/abstract_64258.htm

Scotese, Christopher, 2004. The Devonian Was the Age of Fish! http://www.scotese.com/newpage3.htm

Selden, P. A., 2001. Terrestrialization of Animals. in Paleobiology II, Briggs, D. and Crowther, P. (eds), Blackwell Science, Oxford, UK.

Selover, Robert W. et al., 2005. An Estuarine Assemblage from the Middle Devonian Trout Valley Formation of Northern Maine. Palaios 20(2): 192-197.

Society of Vertebrate Paleontology, 2007. The Devonian, http://www.paleoportal.org/index.php?globalnav=time_space&sectionnav=period&period_id=13

Speer, Brian R., 1998a. Life of the Devonian, http://www.ucmp.berkeley.edu/devonian/devlife.html

—, 1998b, The Devonian, http://www.ucmp.berkeley.edu/devonian/devonian.html

Stein, William E. et al., 2007. Giant Cladoxylopsid Trees Resolve the Enigma of the Earth's Earliest Forest Stumps at Gilboa. Nature v. 446, April 19, pp. 904-907.

Stewart, Wilson N. and Gar W. Rothwell, 1993. Paleobotany and the Evolution of Plants, 2nd ed. Cambridge University Press, Cambridge, UK.

Stock, Carl W., 2003. Originations and Extinctions of Stromatoporoid Genera and Their Role in the Frasnian-Famennian Extinction. http://gsa.confex.com/gsa/2003AM/finalprogram/abstract_61664.htm

University of Rochester, 2007. Rugosan Corals, http://www.earth.rochester.edu/ees207/oct28.html

Wang, Chunzeng, 2007. Presque Isle Geology Field Trip Guide, http://www.umpi.maine.edu/~wangc/field_trips/presque_isle_trip_guide.htm

Whiteley, Thomas E., Gerald J. Kloc and Carlton E. Brett, 2002. Trilobites of New York. Comstock Publishing, Ithaca, NY.

Yeager, Kevin M., 1996. Fossil Fishes (Arthrodira and Acanthodia) from the Upper Devonian Chadakoin Formation of Erie County, Pennsylvania. Ohio Journal of Science 96(3): 52-56.

Ch. 6: Mississippian

Anonymous, 2004a. Mississippian Fossils of Missouri, http://www.lakeneosho.org/Mississippian.html

—, 2004b. The Mississippian, http://www.palaeos.com/Paleozoic/Carboniferous/Mississippian.htm

—, 2007a. Grand Canyon Rock Layers, http://www.bobspixels.com/kaibab.org/geology/gc_layer.htm

—, 2007b. High Inyo Mountain Fossils, http://members.aol.com/Waucoba7/im/inyofossils.html

Barton, Kate E. et al., 2003. The North America Tapestry of Time and Terrain, http://www.tapestry.usgs.gov

Brigham Young University, 2004. Rock Canyon, Utah, http://geologyindy.byu.edu/faculty/rah/slides/Rock%20Canyon/Mississippian%20Sea/Miss%20Main.htm

Butts, Susan H., 2007. Silicified Carboniferous (Chesterian) Brachiopoda of the Arco Hills Formation, Idaho. Journal of Paleontology 81(1): 48-63.

Carroll, Robert L., 1988. Vertebrate Paleontology and Evolution. W. H. Freeman Co., New York, NY.

Chaper, Joseph, 2004. Geology of Missouri—Mississippian, http://members.socket.net/~joschaper/misp.html

Christensen, Ann M., 1999. Brachiopod Paleontology and Paleoecology of the Lower Mississippian Lodgepole Limestone in Southeastern Idaho. pp. 57-67 in Hughes, S. S. and Thackray, G. D. (eds), Guidebook to the Geology of Eastern Idaho, Idaho Museum of Natural History, Pocatello, ID.

Coates, M. I. 2001. Origin of tetrapods. in Paleobiology II, Briggs, D. and Crowther, P. (eds), Blackwell Science, Oxford, UK.

Crowder, David L. and R. Keith Crowder, 1991. Measured Sections and Environmental Interpretations of the Endicott Group (Mississippian), Northeastern Arctic National Wildlife Refuge, Alaska. Alaska Division of Geological and Geophysical Surveys public data file 91-11.

DiMichele, W. A., 2001. Carboniferous coal-swamp forests. in Paleobiology II, Briggs, D. and Crowther, P. (eds), Blackwell Science, Oxford, UK.

— and Robert Hook (rapporteurs), 1992. Paleozoic terrestrial ecosystems. in Behrensmeyer, Anna K. et al. (eds), Terrestrial Ecosystems Through Time. University of Chicago Press, Chicago, IL.

Falls of the Ohio State Park website, 2007, http://www.fallsoftheohio.org

Fichter, Lynn S. and Steven J. Baedke, 2000. The Geological Evolution of Virginia and the Mid-Atlantic Region: the Mississippian Orogenic Calm, http://csmres.jmu.edu/geollab/vageol/vahist/Ja-Miss.html

Gore, Pamela J. W., 1999. The Carboniferous. http://www.dc.peachnet.edu/~pgore/geology/geo102/carbonif.htm

Hagadorn, James W., 2002. Bear Gulch: an exceptional Upper Carboniferous plattenkalk. in Bottjer, David J. et al. (eds), Exceptional Fossil Preservation. Columbia University Press, New York.

Hansen, Michael C., 2001. The Geology of Ohio—the Mississippian. http://www.ohiodnr.com/geosurvey/oh_geol/O1_no_2/missa.htm

Haywick, Douglas, 1998. Thin Section Photomicrographs from the Mississippian Bangor Formation, North Alabama, http://home.earthlink.net/~dragontea/missimag.html

Hoare, Richard D., 2001. Early Mississippian Polyplacophora (Mollusca) from Iowa. Journal of Paleontology 75(1): 66-74.

Hoe, Angela et al., 1998a. Life of the Carboniferous. http://www.ucmp.berkeley.edu/carboniferous/carblife.html

—, 1998b. The Carboniferous. http://www.ucmp.berkeley.edu/carboniferous/carboniferous.html

Humboldt University Natural History Museum, 2004. Mississippian, http://www.humboldt.edu/~natmus/Exhibits/Life_time/Mississippian.web/

Idaho State University, 2007. Fossils in Idaho, http://imnh.isu.edu/digitalatlas/geo/fossils/fossils.htm

Johnson, Thomas, 2004. House of Phacops: the Mississippian, Pennsylvanian, and Permian Periods, http://www.phacops.com/otherperiods.html

Kerp, Hans, 2000. A History of Paleozoic forests, part 2: the Carboniferous coal swamp forests. http://www.biologie.uni-hamburg.de/b-online/kerp/ewald1.html

Lochhaas, David, 2004. *Zaphrentis* - Mississippian Horn Coral, http://www.midamerica.net/~lochhaas/page12.html

McDonald, Karen, 2001. A Paleoecological Reconstruction of the Imo Formation from the Mississippian Era at the Peyton Creek Road Cut in North Central Arkansas, http://www.uca.edu/divisions/academic/biology/students/ksm/poster2001

— and Ben Waggoner, 2000. Ophiuroids from the Imo Formation (Chesterian: Mississippian) of Northern Arkansas. http://www.geosociety.org/absindex/annual/2000/51492.htm

McElwain, J. C., 2001. Atmospheric carbon dioxide—stomata. in Paleobiology II, Briggs, D. and Crowther, P. (eds), Blackwell Science, Oxford, UK.

Miller, Barry W., 2004. A Mississippian and Pennsylvanian Regional Correlation Study on the Pine Mountain Overthrust Area of Tennessee and Part of Kentucky, http://gsa.confex.com/gsa/2004NE/finalprogram/abstract_70966.htm

Northern Arizona University, 2004a. Early Mississippian 340 Ma, http://jan.ucc.nau.edu/~rcb7/Miss.jpg

—, 2004b. Mississippian Paleogeography, Southwestern US, http://jan.ucc.nau.edu/~rcb7/mispaleo.html

Nova Scotia Museum of Natural History, 2004. Fossils of Nova Scotia: Lower Carboniferous (Mississippian) Period, http://museum.gov.ns.ca/fossils/gallery/miss.htm

Ogg, James, 2003. Overview of Global Boundary Stratotype Sections and Points (GSSP's). http://www.stratigraphy.org/gssp.htm

Rohde, Robert, 2004. GeoWhen Database: Mississippian Epoch. http://www.stratigraphy.org/geowhen/stages/Mississippian.html

Sam Noble Museum of Natural History, 2004. Common Fossils of Oklahoma: Mississippian, http://www.snomnh.ou.edu/COLLECTIONS&RESEARCH/sub/common_fossils_of_ok/Mississippian1.htm

Scotese, Christopher, 2000. During the Early Carboniferous Pangea Begins to Form. http://www.scotese.com/newpage4.html

Stanton, Robert J., David L. Jeffer, and Wayne M. Ahr, 2002. Early Mississippian climate based on oxygen isotope compositions of brachiopods, Alamogordo Member of the Lake Valley Formation, south-central New Mexico. Geological Society of America Bulletin 114(1): pp. 4-11.

Svitil, Kathy A., 2003. Mississippian Monsters. Discover Magazine Online, http://www.discover.com/web-exclusives-archive/mississippian-monsters1125/

University of Kentucky, 2004a. Mississippian Age in Kentucky, http://www.uky.edu/KGS/coal/webfossl/pages/missippi.htm

—, 2004b. Strata of Mississippian Age, http://www.uky.edu/KGS/coal/webgeoky/pages/mississippian.html

Wolf, Robert C., 2006. Fossils of Iowa: Field Guide to Paleozoic Deposits. Backinprint.com edition, Lincoln, NE.

Wozniak, Carl, 2004. The Mississippian, http://www.carlwozniak.com/earth/Mississi.html

Ch. 7: Pennsylvanian

Anonymous, 2004. Marine Fossils from Pennsylvanian Deposits of Ferguson, Missouri, http://www.usgennet.org/usa/mo/county/stlouis/ferguson1.html

—, 2004. Pennsylvanian Fossils of Arizona, http://www.psiaz.com/Schur/azpaleo/pennsylv.html

Barton, Kate E. et al., 2003. The North America Tapestry of Time and Terrain, http://www.tapestry.usgs.gov

Behrendt, Marc, 2001a. Feature fossil: dragonfly. Fossil News 7(7): p. 3.

—, 2001b. Feature fossil: Lepidodendron. Fossil News 7(8): p. 3.

Bennington, J. Bret, 1996. Stratigraphic and biofacies patterns in the Middle Pennsylvanian Magoffin marine unit in the Appalachian Basin, USA. International Journal of Coal Geology 31: 169-194.

Carroll, Robert L., 1988. Vertebrate Paleontology and Evolution. W. H. Freeman Co., New York, NY.

Chaper, Joseph, 2004. Geology of Missouri—Pennsylvanian. http://members.socket.net/~joschaper/penn.html

Chestnut, Don, 2000. Pennsylvanian Period, http://www.uky.edu/KGS/coal/webfossl/pages/pennsyl.htm

Clos, Lynne M., 2003. Field Adventures in Paleontology. Fossil News, Boulder, CO.

DiMichele, W. A., 2001. Carboniferous coal-swamp forests. in Paleobiology II, Briggs, D. and Crowther, P. (eds), Blackwell Science, Oxford, UK.

DiMichele, William and Robert Hook (rapporteurs), 1992. Paleozoic terrestrial ecosystems. in Behrensmeyer, Anna K. et al. (eds), Terrestrial Ecosystems Through Time. University of Chicago Press, Chicago, IL.

Gritzo, Ludwig, Lois Gritzo, and LeGrand Smith, 2000. Pennsylvanian Invertebrate Fauna from New Mexico: Brachiopods. Fossil News 6(3): 4 7.

Harris, Ann G., 2003. Geology of National Parks, 6th edition. Kendall-Hunt Publishing, Iowa.

Henson, Harvey, and Phillip Szymcek et al., 2000a. Pennsylvanian biodiversity evidence from the Energy Shale at Carterville, Illinois, http://www.science.siu.edu/geology/big/paper.htm

—, 2000b. Basics in Geology: Pennsylvanian Fossil Study. http://www.science.siu.edu/geology/big/fossils.htm

Hieb, Monte, 1998a. Plant Fossils of West Virginia: Fossil Plants of the Middle Pennsylvanian Period, http://www.clearlight.com/~mhieb/WVFossils/Lycopods.html

—, 1998b. Pennsylvanian Period, http://www.clearlight.com/~mhieb/WVFossils/PennsylvPeriod.html

Hoe, Angela, Azalea Jusay, Ray Mayberry, and Connie Yu, 1998a. The Carboniferous, http://www.ucmp.berkeley.edu/carboniferous/carboniferous.html

—, 1998b. Life of the Carboniferous, http://www.ucmp.berkeley.edu/carboniferous/carblife.html

—, 1998c. Localities of the Carboniferous: Mazon Creek. http://www.ucmp.berkeley.edu/carboniferous/mazon.html

Houck, Karen and Martin Lockley, 1986. Pennsylvanian Biofacies of the Central Colorado Trough. University of Colorado at Denver Geology Department Magazine special issue #2.

Illinois State Museum, 2004. About the Mazon Creek Fossils and Deposit, http://www.museum.state.il.us/exhibits/mazon_creek/about_mazon_creek.html

Itano, Wayne, 2002. Minturn Formation Fossils, http://www.itano.net/fossils/minturn/minturn.htm

—, Karen Houck, and Martin Lockley, 2002. Chondrichthyans from the Pennsylvanian Minturn Formation of Colorado. http://www.itano.net/fossils/projects/earlyvert1.pdf

Johnson, Kirk, and Richard Stucky, 1995. Prehistoric Journey: a History of Life on Earth. Roberts Rinehart Publishers, Boulder, CO.

Kentucky Paleontological Society, 2004. Hazard, KY Field Trip, http://www.uky.edu/OtherOrgs/KPS/pages/hazard.html

Kerp, Hans, 2000a. A History of Paleozoic forests, part 2: the Carboniferous coal swamp forests. http://www.biologie.uni-hamburg.de/b-online/kerp/ewald1.html

—, 2000b. A History of Paleozoic forests, part 3: the Floral Change at the End of the Westphalian. http://www.biologie.uni-hamburg.de/b-online/kerp/ewald2.html

Konecny, James, 2000. The Mazon Creek Nodules. Fossil News 6(9): 4-7.

Labandeira, C. C., 2001. Rise and Diversification of Insects. in Paleobiology II, Briggs, D. and Crowther, P. (eds), Blackwell Science, Oxford, UK.

National Park Service, 2007. Honaker Trail Formation, http://3dparks.wr.usgs.gov/coloradoplateau/lexicon/honaker.htm

Nova Scotia Museum, 2004. Fossils of Nova Scotia: Upper Carboniferous (Pennsylvanian) Period, http://museum.gov.ns.ca/fossils/gallery/penn.htm

Ogg, James, 2003. Overview of Global Boundary Stratotype Sections and Points (GSSP's). http://www.stratigraphy.org/gssp.htm

Pabian, Roger, 2004. Nebraska's Invertebrate Fossils, http://csd/unl.edu/fossils/nebrinbert.asp

Peterson, Morris S., J. Keith Rigby, and Lehi F. Hintze, 1973. Historical Geology of North America. Wm. C. Brown Co., Dubuque, IA.

Ribokas, Bob, 2000. Grand Canyon Rock Layers, http://www.bobspixels.com/kaibab.org/geology/gc_layer.htm

Rice, Charles L., 2001. Contributions to the Geology of Kentucky: Pennsylvanian System, http://pubs.usgs.gov/prof/p1151h/penn.html

Sam Noble Oklahoma Museum of Natural History, 2004. Common Fossils of Oklahoma—Pennsylvanian. http://www.snomnh.ou.edu/COLLECTIONS&RESEARCH/sub/common_fossils_of_ok/Pennsylvanian1.htm

Schellenberg, Stephen A., 2002. Mazon Creek: Preservation in Late Paleozoic Deltaic and Marginal Marine Environments. in Bottjer, David J. et al. (eds), Exceptional Fossil Preservation. Columbia University Press, New York.

Slamen, J., 2003. Pennsylvanian Fossils of North Texas (review), http://www.hgms.org/Articles/PFNTx.html

Speer, Brian R., 1998. Localities of the Carboniferous: Dendrerpeton and Joggins, Nova Scotia, http://www.ucmp.berkeley.edu/carboniferous/joggins.html

Stewart, Wilson N. and Gar W. Rothwell, 1993. Paleobotany and the Evolution of Plants, 2nd ed. Cambridge University Press, Cambridge, UK.

Sutton, Barry G., 2004. Pennsylvanian Fossils of Missouri, http://www.lakeneosho.org/Lake.Neosho.html

Tabor, Neil J., et al., 2001. Stable isotope paleoclimate reconstruction from Permo-Pennsylvanian paleosols of the southwestern United States, http://gsa.confex.com/gsa/2001ESP/finalprogram/abstract_7639.htm

University of Kansas, 2004. Common Fossils of Kansas—Pennsylvanian Plants. http://www.kgs.ukans.edu/Publications/ancien/f05_plants.html

Ch. 8: Permian

Alexander, Chavé, Henry Chang, Carl Tsai, and Peggy Wu, 1998a. The Permian, http://www.ucmp.berkeley.edu/permian/permian.html

—, 1998b. Localities of the Permian: Glass Mountains, Texas, http://www.ucmp.berkeley.edu/permian/glassmts.html

Anonymous, 2004a. Speculated Causes of the Permian Extinction, http://hannover.park.org/Canada/Museum/extinction/permcause.html

—, 2004b. The Permian Mass Extinction, http://hannover.park.org/Canada/Museum/extinction/permass.html

—, 2004c. The Permian-Triassic Mass Extinction, http://www.geocities.com/earthhistory/permo.htm

—, 2004d. Recovery from the Permo-Triassic Filter. http://palaeo.gly.bris.ac.uk/Palaeofiles/Permian/survivors.html

Arens, Nan C. and Brian R. Speer, 1999. Cycads: fossil record, http://www.ucmp.berkeley.edu/seedplants/cycadophyta/cycadfr.html

Ball, Philip, 2001. Brimstone Pickled Permian, Nature Online, http://www.nature.com/nsu/010920/010920-6.html

Barton, Kate E. et al., 2003. The North America Tapestry of Time and Terrain, http://www.tapestry.usgs.gov

Beckemeyer, Roy J., 2004. Fossil insects, http://www.windsofkansas.com/fossil_insects.html

Behrendt, Marc, 2002. Feature fossil: Labidosaurus hamatus. Fossil News 8 (8): p. 2.

Carroll, Robert L., 1988. Vertebrate Paleontology and Evolution. W. H. Freeman Co., New York, NY.

Colbert, Edwin H., 1980. Evolution of the Vertebrates. John Wiley & Sons, NY.

Deer, W. A., R. A. Howie, and J. Zussman, 1966. An Introduction to the Rock-Forming Minerals. Longman Publications, Burnt Mill, England.

DiMichele, William and Robert Hook, 1992. Paleozoic terrestrial ecosystems. in Behrensmeyer, Anna K. et al. (eds), <u>Terrestrial Ecosystems Through Time</u>. University of Chicago Press, Chicago, IL.

Goodwin, Anna, Jon Wyles, and Alex Morley, 2001. The Permo-Triassic Extinction, http://palaeo.gly.bris.ac.uk/Palaeofiles/Permian/intro.html (and links thereon)

Gore, Pamela J. W., 1999. The Permian Period, http://www.dc.peachnet.edu/~pgore/geology/geo102/permian.htm

Hopson, J. A., 2001. Origin of Mammals. in <u>Paleobiology II</u>, Briggs, D. and Crowther, P. (eds), Blackwell Science, Oxford, UK.

Humboldt State University, 2004. Permian, www.humboldt.edu/~natmus/Exhibits/Life_time/Permian_Triassic.web

Idaho State University, 2007. Fossils in Idaho, http://imnh.isu.edu/digitalatlas/geo/fossils/fossils.htm

Ivanov, B. A. and H. J. Melosh, 2003. Eruptions close to the crater. Geology 31 (10): 869-872.

Johnson, Kirk, and Richard Stucky, 1995. <u>Prehistoric Journey: a History of Life on Earth</u>. Roberts Rinehart Publishers, Boulder, CO.

King, Philip B., 1977. <u>The Evolution of North America</u>. Princeton University Press, Princeton, NJ.

LeBeau, Kara, 2001. Permian impact caused largest mass extinction on earth, http://www.spacedaily.com/news/life-01ze.html

McKee, Edwin D., 1931. <u>Ancient Landscapes of the Grand Canyon Region</u>. Northland Press, Flagstaff, AZ.

Mierzejewske, Piotr, 2007. The Hemichordate Class Graptolithoidea, http://www.graptolite.net/Graptolithoidea.html

Ogg, James, 2003. Overview of Global Boundary Stratotype Sections and Points (GSSP's). http://www.stratigraphy.org/gssp.htm

Peterson, Morris S., J. Keith Rigby, and Lehi F. Hintze, 1973. <u>Historical Geology of North America</u>. Wm. C. Brown Co., Dubuque, IA.

Ribokas, Bob, 2000. Grand Canyon Rock Layers, http://www.bobspixels.com/kaibab.org/geology/gc_layer.htm

Santucci, Vince, 2007. Glen Canyon NRA & Rainbow Bridge NM GRI Workshop Summary, http://www.nature.nps.gov/grd/geology/paleo/yell.pdf

— and V. Luke Santucci, Jr., 2004. <u>An Inventory of Paleontological Resources from Walnut Canyon National Monument, Arizona</u>. National Park Service, Kemmerer, WY.

Scotese, Christopher, 2000a. At the end of the Permian was the greatest extinction of all time, http://www.scotese.com/newpage5.htm

—, 2000b. Late & Middle Permian climate, http://www.scotese.com/lpermcli.htm

Speer, Brian R., 1995. Seed plants: fossil record, http://www.ucmp.berkeley.edu/seedplants/seedplantsfr.html

Stewart, Wilson N. and Gar W. Rothwell, 1993. <u>Paleobotany and the Evolution of Plants, 2nd ed</u>. Cambridge University Press, Cambridge, UK

University of Chicago, 2004. Permian introduction, http://pgap.uchicago.edu/Permintro2.html

WGBH-Boston, 2002. Permian-Triassic Extinction, http://www.pbs.org/wgbh/evolution/library/03/2/1_032_02.html

Wignall, P. B., 2001. End-Permian Extinction. in <u>Paleobiology II</u>, Briggs, D. and Crowther, P. (eds), Blackwell Science, Oxford, UK.

Yale Peabody Museum, 2004. Permian, http://www.peabody.yale.edu/mural/permian

Ch. 9: Triassic

Anonymous, 2004a. Bolide impacts, http://palaeo.gly.bris.ac.uk/Palaeofiles/Triassic/bolide.htm

—, 2004b. Ecology of the Triassic, http://palaeo.gly.bris.ac.uk/Palaeofiles/Triassic/ecoloftri.htm

—, 2004c. Mass extinction events caused by a fluctuating sea level, http://palaeo.gly.bris.ac.uk/Palaeofiles/Triassic/sealevel.htm

—, 2004d. The effect of volcanism, http://palaeo.gly.bris.ac.uk/Palaeofiles/Triassic/vulc.htm#The%20Effect%20of%20Volcanism:

—, 2004e. Triassic Paleogeography, Southwestern US, http://jan.ucc.nau.edu/~rcb7/tripaleo.html

—, 2007a. Middle Triassic Ammonoid Fossils from Nevada, http://members.aol.com/Waucoba7/ammonoids/ammonoids.html

—, 2007b. Ammonoids at Union Wash, California, http://members.aol.com/Waucoba7/uw/uwfieldtrip.html

Asaravala, Manish et al., 2000. Triassic Period: tectonics and paleoclimate, http://www.ucmp.berkeley.edu/mesozoic/triassic/triassictect.html

Barton, Kate E. et al., 2003. The North America Tapestry of Time and Terrain, http://www.tapestry.usgs.gov

Brusatte, Stephen, 2002. Stately Fossils. Fossil News, Boulder, CO.

Carroll, Robert L., 1988. Vertebrate Paleontology and Evolution. W. H. Freeman Co., New York, NY.

Clos, Lynne M., 2003. Field Adventures in Paleontology. Fossil News, Boulder, CO.

Coffey, Brian P. and Daniel A. Textoris, 2003. Paleosols and Paleoclimate Evolution. in The Great Rift Valleys of Pangea in Eastern North America (vol. 2: Sedimentology, Stratigraphy, and Paleontology), Peter M. LeTourneau and Paul E. Olsen (eds), Columbia University Press, New York, NY.

Cornet, Bruce and Alfred Traverse, 1973. Fossil Spores, Pollen, and Fishes from Connecticut Indicate Early Jurassic Age for Part of the Newark Group. Science 182, pp. 1243-1247.

Etter, Walter, 2002. Grès à Voltzia: Preservation in Early Mesozoic Deltaic and Marginal Marine Environments. in Bottjer, D. et al (eds), Exceptional Fossil Preservation. Columbia University Press, New York, NY.

Fedak, Tim J., 2003. Erosion of Late Triassic Sandstones (Carr's Brook Formation) on the Northern Shore of the Minas Basin Provide New Vertebrate Fossil Specimens, GSA Northeastern Section general meeting paper no. 34-4, http://gsa.confex.com/gsa/2003NE/finalprogram/abstract_51197.htm

Glut, Donald F., 2003. Dinosaurs, the Encyclopedia, Supplement 3. McFarland & Co., Jefferson, NC.

Hayden, Martha, 2004. Dinosaurs of Utah, http://www.media.utah.edu/UHE/d/DINOSAURSOFUT.html

Illinois State Geological Survey, 2004. Triassic Period, http://www.isgs.uiuc.edu/dinos/de_4/5c5cae8.htm

Johnson, Kirk, and Richard Stucky, 1995. Prehistoric Journey: a History of Life on Earth. Roberts Rinehart Publishers, Boulder, CO.

Kent, Dennis V. and Giovanni Muttoni, 2003. Mobility of Pangea: Implications for Late Paleozoic and Early Mesozoic Paleoclimate. in The Great Rift Valleys of Pangea in Eastern North America (vol. 1: Tectonics, Structure, and Volcanism), Peter M. LeTourneau and Paul E. Olsen (eds), Columbia University Press, New York, NY.

Long, Robert A. and Rose Houk, 1988. Dawn of the Dinosaurs: the Triassic in the Petrified Forest. Petrified Forest Museum Association, Petrified Forest, AZ.

McHone, J. Gregory and John H. Puffer, 2003. Flood Basalt Provinces of the Pangean Atlantic Rift: Regional Extent and Environmental Significance. in The Great Rift Valleys of Pangea in Eastern North America (vol. 1: Tectonics, Structure, and Volcanism), Peter M. LeTourneau and Paul E. Olsen (eds), Columbia University Press, New York, NY.

Ogg, James, 2003. Overview of Global Boundary Stratotype Sections and Points (GSSP's). http://www.stratigraphy.org/gssp.htm

Olsen, Paul E., 2000. Great Triassic Assemblages, part 2: the Fleming Fjord Formation, http://www.ideo.columbia.edu/edu/dees/courses/v1001/keuper11.html

Olsen, Paul E. and J. Gregory McHone, 2003. Introduction to the Central Atlantic Large Igneous Province. in The Great Rift Valleys of Pangea in Eastern North America (vol. 1: Tectonics, Structure, and Volcanism), Peter M. LeTourneau and Paul E. Olsen (eds), Columbia University Press, New York, NY.

Parker, Steve, 2003. Dinosaurus: the Complete Guide to Dinosaurs. Firefly Books, Buffalo, NY.

Peterson, Morris S., J. Keith Rigby, and Lehi F. Hintze, 1973. Historical Geology of North America. Wm. C. Brown Co., Dubuque, IA.

Russell, Dale A., 1989. An Odyssey in Time: the Dinosaurs of North America. North Word Press, Minocqua, WI.

Rowland, Stephen M., 1999. The Ichthyosaur, Nevada's State Fossil. Rocks and Minerals, November 1999.

Scotese, Christopher, 2000. At the end of the Triassic, Pangea began to rift apart, http://www.scotese.com/newpage8.htm

Speer, Brian R., 1997. The Triassic Period, http://www.ucmp.berkeley.edu/mesozoic/triassic/triassic.html

Stewart, Wilson N. and Gar W. Rothwell, 1993. Paleobotany and the Evolution of Plants, 2nd ed. Cambridge University Press, Cambridge, UK.

Szajna, Michael J. and Brian W. Hartline, 2003. A New Vertebrate Footprint Locality from the Late Triassic Passaic Formation Near Birdsboro, Pennsylvania. in The Great Rift Valleys of Pangea in Eastern North America (vol. 2: Sedimentology, Stratigraphy, and Paleontology), Peter M. LeTourneau and Paul E. Olsen (eds), Columbia University Press, New York, NY.

Taylor, M. A., 2001. Locomotion in Mesozoic Marine Reptiles. in Paleobiology II, Briggs, D. and Crowther, P. (eds), Blackwell Science, Oxford, UK.

Wing, Scott L. and Hans-Dieter Sues (rapporteurs), 1992. Mesozoic and Early Cenozoic Terrestrial Ecosystems. in Behrensmeyer, Anna K. et al. (eds), Terrestrial Ecosystems Through Time. University of Chicago Press, Chicago, IL.

Ch. 10 - Jurassic

Anonymous, 2004. North American Dinosaurs, http://www.isgs.uiuc.edu/dinos/de_4/5c60e18.htm

Armstrong, Wayne, 2000. Plants of Jurassic Park, http://waynesword.palomar.edu/ww0803.htm

Barnes, Fran, 2004. Navajo Sandstone Fossils, http://www.lakepowell.net/navajowet.html

Barton, Kate E. et al., 2003. The North America Tapestry of Time and Terrain, http://www.tapestry.usgs.gov

Brusatte, Stephen, 2002. Stately Fossils. Fossil News, Boulder, CO.

Bureau of Land Management, 2007a. Red Gulch Dinosaur Tracksite, http://www.blm.gov/wy/st/en/field_offices/Worland/Tracksite.html

—, 2007b. Cleveland-Lloyd Dinosaur Quarry, http://www.blm.gov/utah/price/quarry.html

Carroll, Robert L., 1988. Vertebrate Paleontology and Evolution. W. H. Freeman Co., New York, NY.

Chiappe, Luis M., 2001. The Rise of Birds. in Paleobiology II, Briggs, D. and Crowther, P. (eds), Blackwell Science, Oxford, UK.

—, 2007. Glorified Dinosaurs: the Origin and Early Evolution of Birds. John Wiley & Sons, Hoboken, NJ.

Columbia University, 2004. The Real Jurassic Park - Morrison and Tendaguru Formations, http://rainbow.ldeo.columbia.edu/courses/v1001/14.html

Etter, Walter, 2002. Solnhofen: Plattenkalk Preservation with *Archaeopteryx*. in Bottjer, D. et al (eds), Exceptional Fossil Preservation. Columbia University Press, New York, NY.

— and Carol M. Tang, 2002. Posidonia Shale: Germany's Jurassic Marine Park. in Bottjer, D. et al (eds), Exceptional Fossil Preservation. Columbia University Press, New York, NY.

Farlow, James O. and Peter M. Galton, 2003. Dinosaur Trackways of Dinosaur State Park, Rocky Hill, CT. in The Great Rift Valleys of Pangea in Eastern North America (vol. 2: Sedimentology, Stratigraphy, and Paleontology), Peter M. LeTourneau and Paul E. Olsen (eds), Columbia University Press, New York, NY.

Foster, John, 2003. Tracking in Grand Staircase-Escalante National Monumeent. Fossil News 9(10): 4-9.

Gishlick, A. D., 2001. Predatory Behavior in Maniraptoran Theropods. in Paleobiology II, Briggs, D. and Crowther, P. (eds), Blackwell Science, Oxford, UK.

Hamblin, Alden H., 2002. Middle Jurassic Fossil Footprints in the San Rafael Swell, Utah and Their Affinity to Tracks in the Grand Staircase-Escalante National Monument and Elsewhere in Utah, http://gsa.confex.com/gsa/2002RM/finalprogram/abstract_34238.htm

Harris, Jerald D. and Kenneth A. Lacovaara, 2004. Enigmatic Fossil Footprints from the Sundance Formation (Upper Jurassic) of Bighorn Canyon National Recreation Area, Wyoming. Ichnos 11: 151-166.

Kielan-Jaworowska, Zofia, Richard L. Cifelli, and Zhe-Xi Luo, 2004. Mammals From the Age of Dinosaurs. Columbia University Press, New York.

Kelley, P. H. and T. A. Hansen, 2001. Mesozoic Marine Revolution. in Paleobiology II, Briggs, D. and Crowther, P. (eds), Blackwell Science, Oxford, UK.

Lockley, Martin G., 2001. Trackways—Dinosaur Locomotion. in Paleobiology II, Briggs, D. and Crowther, P. (eds), Blackwell Science, Oxford, UK.

Motani, Ryosuke, 2004. Rulers of the Jurassic Seas. Dinosaurs and Other Monsters, Scientific American special edition, pp. 4-11.

National Park Service, 2007. Dinosaur National Monument home page, http://www.nps.gov/archive/dino/morrison.htm

Ogg, James, 2003. Overview of Global Boundary Stratotype Sections and Points (GSSP's). http://www.stratigraphy.org/gssp.htm

Peterson, Morris S., J. Keith Rigby, and Lehi F. Hintze, 1973. Historical Geology of North America. Wm. C. Brown Co., Dubuque, IA.

Prum, Richard O. and Alan H. Brush, 2004. Which Came First, the Feather or the Bird? Dinosaurs and Other Monsters, Scientific American special edition, pp. 72-81.

Russell, Dale A., 1989. An Odyssey in Time: the Dinosaurs of North America. North Word Press, Minocqua, WI.

Scotese, Christopher R., 2004. Early Jurassic, http://www.scotese.com/jurassic.htm

Smith, Joshua B. and James O. Farlow, 2003. Osteometric Approaches to Trackmaker Assignment for the Newark Supergroup Ichnogenera *Grallator*, *Anchisauripus*, and *Eubrontes*. in The Great Rift Valleys of Pangea in Eastern North America (vol. 2: Sedimentology, Stratigraphy, and Paleontology), Peter M. LeTourneau and Paul E. Olsen (eds), Columbia University Press, New York, NY.

Tang, Carol M., 2002. Oxford Clay: England's Jurassic Marine Park. in Bottjer, D. et al (eds), <u>Exceptional Fossil Preservation</u>. Columbia University Press, New York, NY.

Tharp, Stephanie, 2004. *Stegosaurus*, http://www.uky.edu/ArtsSciences/Geology/webdogs/time/jurassic/jura5.htm

Turner, Christine E. and Fred Peterson, 2004. Late Jurassic Ecosystem Reconstruction During Deposition of the Morrison Formation and Related Beds in the Western Interior of the United States, http://www2.nature.nps.gov/geology/publications/hot_topic/ht_morrison.htm

University of Chicago, 2004. Jurassic Floras and Climates, http://pgap.uchicago.edu/JURfossilplants.html

Utah Geological Survey, 2004. Utah in the Age of Dinosaurs, http://geology.utah.gov/utahgeo/dinofossil/dinoage.htm

—, 2007. Dinosaur Tracksites and Trails, http://www.ugs.state.ut.us/utahgeo/dinofossil/dinotracks.htm

Utah State Parks, 2007. Red Fleet State Park, http://www.stateparks.utah.gov/park/about.php?id=RFSP

Weishampel, David B. and Luther Young, 1996. <u>Dinosaurs of the East Coast</u>. Johns Hopkins University Press, Baltimore, MD.

Wing, Scott L. and Hans-Dieter Sues (rapporteurs), 1992. Mesozoic and Early Cenozoic Terrestrial Ecosystems. in Behrensmeyer, Anna K. et al. (eds), <u>Terrestrial Ecosystems Through Time</u>. University of Chicago Press, Chicago, IL.

Wright, Jo, 2004. Walking With Dinosaurs: Late Jurassic, http://dsc.discovery.com/stories/dinos/bbc/chronology/152/

Ch. 11: Cretaceous

Anonymous, 2004a. North American Dinosaurs, http://www.isgs.uiuc.edu/dinos/de_4/5c60e18.htm

—, 2004b. The End-Cretaceous Extinction, http://hannover.park.org/Canada/Museum/extinction/cretmass.html

Barton, Kate E. et al., 2003. The North America Tapestry of Time and Terrain, http://www.tapestry.usgs.gov

Baumiller, Tomasz K., Lindsey R. Leighton, and David L. Thompson, 1999. Boreholes in Mississippian Spiriferid Brachiopods and their Implications for Paleozoic Gastropod Drilling. Palaeo 147: 283-289.

Bennington, J. Bret, 2003. Paleontology and Sequence Stratigraphy of the Upper Cretaceous Navesink Formation, New Jersey. Long Island Geologists Field Trip Guide, October 18, 2003.

British Broadcasting Corporation, 2004. The Mass Extinctions: the End Cretaceous Extinction, http://www.bbc.co.uk/education/darwin/exfiles/cretaceous.htm

Carroll, Robert L., 1988. <u>Vertebrate Paleontology and Evolution</u>. W. H. Freeman Co., New York, NY.

Clos, Lynne M., 2001. Cretaceous and Tertiary Cycads of the Western Interior. Fossil News 7(11): 5-8.

Dinosaur Provincial Park website, http://www.cd. gov.ab.ca/parks/dinosaur

Dinosaur Valley State Park website, http://www.tpwd.state.tx.us/park/dinosaur

Foster, John, 2003. Tracking in Grand Staircase-Escalante National Monumeent. Fossil News 9(10): 4-9.

Gore, Pamela J. W., 1999. The Cretaceous Period, http://www.gpc.edu/~pgore/geology/geo102/cretac.htm

Hildebrand, Alan, 2004. Chicxulub Crater, Mexico, and the Cretaceous-Tertiary Boundary, http://dsaing.uqac.uquebec.ca/~mhiggins/MIAC/chicxulub.htm

Hurum, Jørn H., 1995. Reconstruction of the petrosal in Late Cretaceous multituberculates (Mammalia), http://www.2dgf.dk/online/hurum.htm

Itano, Wayne, 2002. Itano Family Fossil Collection, http://www.itano.net/fossils.htm

Kazlev, M. Alan, 2002. The Cretaceous Period, http://www.palaeos.com/Mesozoic/Cretaceous/Cretaceous.htm

Kielan-Jaworowska, Zofia, Richard L. Cifelli, and Zhe-Xi Luo, 2004. <u>Mammals From the Age of Dinosaurs</u>. Columbia University Press, New York.

King, David T., 1997. Late Cretaceous Dinosaurs of the Southeastern United States, http://www.auburn.edu/~kingdat/dinosaur_webpage.htm

Kring, David A., 2004. Chicxulub Impact Event: Understanding the K-T Boundary, http://www.lpl.arizona.edu/SIC/impact_cratering/Chicxulub/Chicx_title.html

Lipka, Thomas R., 1995. Re: Definition of Chicxulub, http://dml.cmnh.org/1995Jul/msg00262.html

MacLeod, Norman, 2004. Mass Extinctions at the Cretaceous-Tertiary (K-T) Boundary, http://www.nhm.ac.uk/science/intro/palaeo/project3/

Montez, Josie and Amanda Turnbull, 2005. EPSCoR Student Research. The Bronto, v. 14, issue 2, Summer/Fall 2005 (newsletter of University of Wyoming Geological Museum).

Norris, R. D., 2001. Impact of K/T Boundary Events on Marine Life. in Paleobiology II, Briggs, D. and Crowther, P. (eds), Blackwell Science, Oxford, UK.

Ogg, James, 2003. Overview of Global Boundary Stratotype Sections and Points (GSSP's). http://www.stratigraphy.org/gssp.htm

Peterson, Morris S., J. Keith Rigby, and Lehi F. Hintze, 1973. Historical Geology of North America. Wm. C. Brown Co., Dubuque, IA.

Ratkevich, Ron, 2003. Paleontology of the Late Cretaceous Fort Crittenden and Salero Formations of Southwestern Arizona. Tucson Mineral & Gem World, Tucson, AZ.

Russell, Dale A., 1989. An Odyssey in Time: the Dinosaurs of North America. North Word Press, Minocqua, WI.

San Diego Natural History Museum, 2004. Geologic Time Line, http://www.sdnhm.org/fieldguide/fossils/timeline.html

Scotese, Christopher R., 2004a. Late Cretaceous Climate, http://www.scotese.com/lcretcli.htm

—, 2004b. More Information About the Cretaceous, http://www.scotese.com/moreinfo12.htm

Smithsonian Institution, 2004. A Blast from the Past!, http://www.nmnh.si.edu/paleo/blast/index.html

U. S. Geological Survey, 2003. NYC Regional Geology: Late Cretaceous Stratigraphic Units of the Coastal Plain, http://3dparks.wr.usgs.gov/nyc/coastalplain/cretaceous.htm

Utah Geological Survey, 2004a. Early Cretaceous Dinosaurs of Utah, http://geology.utah.gov/utahgeo/dinofossil/dinecret.htm

—, 2004b. Late Cretaceous Dinosaurs of Utah, http://geology.utah.gov/utahgeo/dinofossil/dinlcret.htm

Vickers-Rich, Patricia and Thomas H. Rich, 2004. Dinosaurs of the Antarctic. in Dinosaurs and Other Monsters, Scientific American Special Edition.

W. M. Browning Cretaceous Fossil Park website, http://www.boonevillemississippi.com/recreation/fossil.htm

Waggoner, Ben M. and Brian R. Speer, 1998. The Cretaceous Period: Life, http://www.ucmp.berkeley.edu/mesozoic/cretaceous/cretlife.html

Weishampel, David B. and Luther Young, 1996. Dinosaurs of the East Coast. Johns Hopkins University Press, Baltimore, MD.

Willson, X., 2004. The Cretaceous, http://palaeo.gly.bris.ac.uk/communication/Willson/earlycret.html

Wing, Scott L. and Hans-Dieter Sues (rapporteurs), 1992. Mesozoic and Early Cenozoic Terrestrial Ecosystems. in Behrensmeyer, Anna K. et al. (eds), Terrestrial Ecosystems Through Time. University of Chicago Press, Chicago, IL.

Wolfe, J. A. and Dale A. Russell, 2001. Impact of K/T Boundary Events on Terrestrial Life. in Paleobiology II, Briggs, D. and Crowther, P. (eds), Blackwell Science, Oxford, UK.

Ch. 12: Paleocene

Alroy, John, 2007. The Fossil Record of North American Mammals: Evidence for a Paleocene Evolutionary Radiation, http://www.nceas.ucsb.edu/~alroy/Paleocene.html

Anonymous, 2007a. Paleocene, http://en.wikipedia.org/wiki/Paleocene

—, 2007b. Paleocene-Eocene Thermal Maximum, http://en.wikipedia.org/wiki/Paleocene-Eocene_Thermal_Maximum

—, 2007c. Wannagan Creek Site, http://www.answers.com/topic/wannagan-creek-site

Dykes, Trevor, 2007. Mammal Fossils in the Ferris Formation, Wyoming (Upper Cretaceous-Paleocene), http://www.geocities.com/trevor_dykes/hannabasin.htm

Ellis, Beth, Kirk R. Johnson, and Regan E. Dunn, 2007. Evidence for an in situ Early Paleocene Rainforest from Castle Rock, Colorado, http://pubs.gg.uwyo.edu/38-1_ABSTRACTS.htm

Grande, Lance, Li Guo-Qing, and Mark Wilson, 2000. Amia cf. pattersoni from the Paleocene Paskapoo Formation of Alberta. Canadian Journal of Earth Sciences 37(1): 31-37.

Hoganson, John W. and Johnathan Campbell, 2002. Paleontology of Theodore Roosevelt National Park, http://www.nd.gov/ndgs/NDNotes/ndn9_h.htm

Itano, Wayne, 2002. Maryland Paleocene Fossils, http://www.itano.net/fossils/marylan2/marylan2.htm

Janis, Christine M., Katherine M. Scott, and Louis L. Jacobs, 1998. Evolution of Tertiary Mammals of North America. Cambridge University Press, Cambridge, UK.

Janis, Christine M., 2001. Radiation of Tertiary Mammals. in <u>Paleobiology II</u>, Briggs, D. and Crowther, P. (eds), Blackwell Science, Oxford, UK.

Jehle, Martin, 2007. Paleocene Mammals of the World, http://www.paleocene-mammals.de (and internal links therein)

Kihm, Allen J., 2004. Fossil Mammals of the Sentinel Butte Formation (Late Paleocene) of North Dakota, http://gsa.confex.com/gsa/2004RM/finalprogram/abstract_72521.htm

Klosterman, Susan M. et al., 2001. New Paleocene Rhynchonellidae Brachiopods from the Potrerillos Formation, Northeast Mexico, http://gsa.confex.com/gsa/2001AM/finalprogram/abstract_27988.htm

Ogg, James, 2003. Overview of Global Boundary Stratotype Sections and Points (GSSP's). http://www.stratigraphy.org/gssp.htm

Rose, K. D., 2001. Early Primates. in <u>Paleobiology II</u>, Briggs, D. and Crowther, P. (eds), Blackwell Science, Oxford, UK.

Sanders, Albert E., 1998. Paleobiology of the Williamsburg Formation (Paleocene) of South Carolina. Transactions of the American Philosophical Society, new series, v. 88, no. 4, pp. 255-268.

Wing, Scott L. and Hans-Dieter Sues (rapporteurs), 1992. Mesozoic and Early Cenozoic Terrestrial Ecosystems. in Behrensmeyer, Anna K. et al. (eds), <u>Terrestrial Ecosystems Through Time</u>. University of Chicago Press, Chicago, IL.

Ch. 13: Eocene

Anonymous, 2007a. Eocene, http://en.wikipedia.org/wiki/Eocene

—, 2007b. John Day Fossil Beds National Monument, http://en.wikipedia.org/wiki/Clarno_Formation

—, 2007c . BC Parks - Driftwood Canyon Provincial Park, Smithers, http://www.britishcolumbia.com/parks/?id=618

Barton, Kate E. et al., 2003. The North America Tapestry of Time and Terrain, http://www.tapestry.usgs.gov

Bourdon, Jim and Michael Folmer, 2007. Early Eocene Sharks & Rays of Virginia, http://www.elasmo.com/frameMe.html?file=paleo/fauna/va_eoc.html&menu=fauna-alt.html

Brenes, Kaytee et al., 1998. Localities of the Eocene: Green River Formation, http://www.ucmp.berkeley.edu/tertiary/eoc/greenriver.html

Daniels, Frank J., 2006. <u>Ancient Forests: a Closer Look at Fossil Wood</u>. Western Colorado Publishing Co., Grand Junction, CO.

Davies-Vollum, K. Sian and Scott L. Wing, 1998. Sedimentological, Taphonomic, and Climatic Aspects of Eocene Swamp Deposits (Willwood Formation, Bighorn Basin, Wyoming). Palaios 13(1): 28-40.

Heinrich, Paul V., 2001. Primitive Eocene Whales, http://www.intersurf.com/~chalcedony/Basilosaurus1.html

Janis, Christine M., Kathleen M. Scott, and Louis L. Jacobs (eds), 1998. <u>Evolution of Tertiary Mammals of North America</u>. Cambridge University Press, Cambridge, UK.

Kazlev, M. Alan, 2002. The Eocene, http://www.palaeos.com/Cenozoic/Eocene/Eocene.htm

Kohls, David, 2007. Eocene Fossils, http://www.coloradomtn.edu/campus_rfl/staff_rfl/kohls/eocene.shtml

National Park Service, 2004. Fossil Butte National Monument home page, http://www.nps.gov/archive/fobu/expanded/index.htm

—, 2007. Yellowstone Paleo Survey: Localities, http://www2.nature.nps.gov/geology/paleontology/surveys/yell_survey/local.htm

Ogg, James, 2003. Overview of Global Boundary Stratotype Sections and Points (GSSP's). http://www.stratigraphy.org/gssp.htm

Passmore, Susan, 2006. The Accidental Discovery of North America's Most Diverse Eocene Flora. Fossil News 12(8): 4-8.

Robb, Jim, 1998. The Chesapeake Bay Bolide Impact: a New View of Coastal Plain Evolution, http://marine.usgs.gov/fact-sheets/fs49-98/

Scotese, Christopher, 2007. During the Early Cenozoic India Began to Collide with Asia, http://www.scotese.com/newpage9.htm

Shaw, Bernita L., 2006. The McAbee Fossil Beds, http://www.dll-fossils.com

Turnbull, William D., 2004. Taeniodonta of the Washakie Formation, Southwestern Wyoming. Bulletin of the Carnegie Museum of Natural History 36(1): 303-333.

Waggoner, Ben, 2000. Localities of the Eocene: Florissant Formation, http://www.ucmp.berkeley.edu/tertiary/eoc/florissant.html

Wing, Scott L. and Hans-Dieter Sues (rapporteurs), 1992. Mesozoic and Early Cenozoic Terrestrial Ecosystems. in Behrensmeyer, Anna K. et al. (eds), <u>Terrestrial Ecosystems Through Time</u>. University of Chicago Press, Chicago, IL.

Wong, Kate, 2004. The Mammals That Conquered the Seas. in Dinosaurs and Other Monsters, Scientific American special issue v. 14, no. 2.

Ch. 14: Oligocene

Anonymous, 2007a. Fossil Plants and Insects at Bull Run, Nevada, http://members.aol.com/Waucoba7/br/bullrun.html

—, 2007b. The Oligocene, http://www.mnh.si.edu/anthro/HumanOrigins/faq/gt/cenozoic/oligocene.htm

—, 2007c. John Day Fossil Beds National Monument, http://en.wikipedia.org/wiki/John_Day_Formation

—, 2007d. Lowcountry Geologic, http://www.lowcountrygeologic.com/SaleFossils/tabid/90/Default.aspx

Foss, Scott E., 1995. Fossils of the Badlands of South Dakota, http://northern.edu/natsource/earth/fossil1.htm

Heinrich, Paul V., 2001. Oligocene Vertebrate Fossils, http://www.intersurf.com/~chalcedony/Oligofos.html

Janis, Christine M., Kathleen M. Scott, and Louis L. Jacobs (eds), 1998. Evolution of Tertiary Mammals of North America. Cambridge University Press, Cambridge, UK.

National Park Service, 2006. John Day Fossil Beds Paleontology, http://www.nps.gov/joda/naturescience/john-day-fossil-beds-paleontology.htm

—, 2007. Badlands National Park, http://www.nps.gov/badl

Ogg, James, 2003. Overview of Global Boundary Stratotype Sections and Points (GSSP's). http://www.stratigraphy.org/gssp.htm

Potts, Richard and Anna K. Behrensmeyer (rapporteurs), 1992. Late Cenozoic Terrestrial Ecosystems. in Behrensmeyer, Anna K. et al. (eds), Terrestrial Ecosystems Through Time. University of Chicago Press, Chicago, Il..

Scotese, Christopher, 2007. Oligocene Climate, http://www.scotese.com/oligocen.htm

Singleton, Scott, 2001. Fossil Wood of the Oligocene Catahoula Formation, Jasper County, Texas, http://www.hgms.org/StartPageHTMLfiles/PaleoPetrifiedWoodArticles/2001-8-JasperWoodArticle.html

Speer, Brian R., 1995. Oligocene Epoch: Life, http://www.ucmp.berkeley.edu/tertiary/oli/olilife.html

Waggoner, Ben, 2000. Localities of the Oligocene: the Creede Formation, http://www.ucmp.berkeley.edu/tertiary/oli/creede.html

Wong, Kate, 2004. The Mammals That Conquered the Seas. in Dinosaurs and Other Monsters, Scientific American special issue v. 14, no. 2.

Zanazzi, Alessandro et al., 2007. Large Temperature Drop Across the Eocene-Oligocene Transition in Central North America. Nature 445: 639-642.

Ch. 15: Miocene

Anonymous, 2007a. Fossil Leaves and Seeds in West-Central Nevada, http://members.aol.com/Waucoba7/middlegate/middlegate.html

—, 2007b. Days of Discovery in the Ione Basin, Western Foothills of the Sierra Nevada, Amador County, California, http://members.aol.com/Waucoba5/ione/discovery1.html

—, 2007c. Fossil Plants at Buffalo Canyon, Nevada, http://members.aol.com/Waucoba7/bc/bufcanyon.html

—, 2007d. Ashfall Fossil Beds, http://en.wikipedia.org/wiki/Ashfall_Fossil_Beds

—, 2007e. Gray Fossil Site, http://en.wikipedia.org/wiki/Gray_Fossil_Site

—, 2007f. John Day Fossil Beds National Monument, http://en.wikipedia.org/wiki/Mascall_Formation

Bourdon, Jim et al., 2002. Sharktooth Hill: Miocene Sharks & Rays, http://www.elasmo.com/frameMe.html?file=paleo/sth/shrkhill.html&menu=bin/menu_fauna-alt.html

Cerling, T. E., 2001. Evolution of Modern Grasslands and Grazers. in Paleobiology II, Briggs, D. and Crowther, P. (eds), Blackwell Science, Oxford, UK.

Dillhoff, Thomas A., 2004. Fossil Forests of Eastern Washington, http://www.evolvingearth.org/learnearthscience/sciencearticles0501fossilforest.htm

Eastern Tennessee State University, 2007. Gray Fossil Site, http://www.etsu.edu/grayfossilsite/History/

Garibay-Romero, Luis M., Francisco J. Vega, Torrey G. Nyborg, and Igor Hernández, 2003. Some Additions to the Fossil Decapod Crustaceans of the Miocene Tuxpam Formation, Veracruz, Mexico, http://gsa.confex.com/gsa/2003AM/finalprogram/abstract_66868.htm

Itano, Wayne, 1999. Astoria Formation (Oregon Miocene), http://www.itano.net/fossils/astoria/astoria.htm

—, 2001. Maryland Miocene Fossils, http://www.itano.net/fossils/maryland/maryland.htm

Janis, Christine M., Kathleen M. Scott, and Louis L. Jacobs (eds), 1998. <u>Evolution of Tertiary Mammals of North America</u>. Cambridge University Press, Cambridge, UK.

Jefferson, George T. and Lowell Lindsay (eds), 2006. <u>Fossil Treasures of the Anza-Borrego Desert: the Last Seven Million Years</u>. Sunbelt Publications, San Diego, CA.

Kazlev, M. Alan, 2002. The Miocene, http://www.palaeos.com/Cenozoic/Miocene/Miocene.htm

Kohl, Martin M., 2007. Bedrock Geology of the Gray Site & Vicinity, http://www.state.tn.us/environment/tdg/gray/Geology.shtml

McLennan, Jean D., 1971. <u>Miocene Shark's Teeth of Calvert County</u>, Maryland Geological Survey, Baltimore, MD.

National Park Service, 2006a. John Day Fossil Beds Paleontology, http://www.nps.gov/joda/naturescience/john-day-fossil-beds-paleontology.htm

—, 2006b. Agate Fossil Beds National Monument - Outdoor Activities, http://www.nps.gov/agfo/planyourvisit/outdooractivities.htm

Northern Arizona University, 2007. Early Miocene 20 Ma, http://jan.ucc.nau.edu/~rcb7/Mio.jpg

Ogg, James, 2003. Overview of Global Boundary Stratotype Sections and Points (GSSP's). http://www.stratigraphy.org/gssp.htm

Potts, Richard and Anna K. Behrensmeyer (rapporteurs), 1992. Late Cenozoic Terrestrial Ecosystems. in Behrensmeyer, Anna K. et al. (eds), <u>Terrestrial Ecosystems Through Time</u>. University of Chicago Press, Chicago, IL.

Raynolds, Bob, 1999. Rhino Revelation. Fossil News 5(2): 3-5.

Schiebout, Judith A. et al., 2006. Miocene Vertebrate Fossils Recovered from the Pascagoula Formation in Southeastern Louisiana, http://www.searchanddiscovery.net/documents/2006/06086gcags_sec_abs/images/abstract.schiebout.et.al.pdf

Scotese, Christopher R., 2007a. Middle Miocene: the World Assumes a Modern Configuration, http://www.scotese.com/miocene.htm

—, 2007b. Miocene Climate, http://www.scotese.com/miocene1.htm

Speer, Brian R., 1997a. Miocene Epoch: Life, http://www.ucmp.berkeley.edu/tertiary/mio/miolife.html

—, 1997b. Localities of the Miocene: The Monterey Formation of California, http://www.ucmp.berkeley.edu/tertiary/mio/monterey.html

Stinson, Amy L., Mark Izzi, Lauren Mirallegro, and Laura Tewksbury, 2005. Mid-Miocene Cetacean Fossils, Paularino Member, Topanga Formation, Western San Joaquin Hills, Orange County, California, http://gsa.confex.com/gsa/2005CD/finalprogram/abstract_85790.htm

Voorhies, Mike, 1992. Ashfall: Life and Death at a Nebraska Waterhole Ten Million Years Ago. Museum Notes no. 81, University of Nebraska State Museum, Lincoln, NE.

Ch. 16: Pliocene

Anonymous, 2007a. Field Trip to the Kettleman Hills Fossil Field in Kings County, California, http://members.aol.com/Waucoba5/kh/kettlefieldtrip.html

—, 2007b. Fossil Bones in the Coso Range, California, http://members.aol.com/Waucoba5/coso/coso.html

—, 2007c. Pipe Creek Sinkhole, http://en.wikipedia.org/wiki/Pipe_Creek_Sinkhole

—, 2007d. Pliocene, http://en.wikipedia.org/wiki/Pliocene

Bourdon, Jim, 2004. Lee Creek aka Aurora: a Neogene Fauna from North Carolina, http://www.elasmo.com/frameMe.html?file=leecreek/leecreek.html&menu=bin/menu_leecreek-alt.html

Dodd, C. Kenneth and Gary S. Morgan, 1992. Fossil Sea Turtles from the Early Pliocene Bone Valley Formation, Central Florida. Journal of Herpetology 26(1): 1-8.

Emslie, Steven D., 2007. Fossil Passerines from the Early Pliocene of Kansas and the Evolution of Songbirds in North America. The Auk, January.

Heinrich, Paul V., 2001. Pliocene Fossils, http://www.intersurf.com/~chalcedony/Pliofos.html

Itano, Wayne, 2001. San Diego Formation (California Pliocene), http://www.itano.net/fossils/sandiego/sandiego.htm

Janis, Christine M., Kathleen M. Scott, and Louis L. Jacobs (eds), 1998. <u>Evolution of Tertiary Mammals of North America</u>. Cambridge University Press, Cambridge, UK.

Jefferson, George T. and Lowell Lindsay (eds), 2006. Fossil Treasures of the Anza-Borrego Desert: the Last Seven Million Years. Sunbelt Publications, San Diego, CA.

Kazlev, M. Alan, 2002. The Pliocene, http://www.palaeos.com/Cenozoic/Pliocene/Pliocene.htm

National Park Service, 2007. Geology Fieldnotes: Hagerman Fossil Beds National Monument, Idaho, http://www2.nature.nps.gov/geology/parks/hafo/#geology

Ogg, James, 2003. Overview of Global Boundary Stratotype Sections and Points (GSSP's). http://www.stratigraphy.org/gssp.htm

Potts, Richard and Anna K. Behrensmeyer (rapporteurs), 1992. Late Cenozoic Terrestrial Ecosystems. in Behrensmeyer, Anna K. et al. (eds), Terrestrial Ecosystems Through Time. University of Chicago Press, Chicago, IL.

Ruez, Dennis R., 2006. Middle Pliocene Paleoclimate in the Glenns Ferry Formation of Hagerman Fossil Beds National Monument, Idaho: a Baseline for Evaluating Faunal Change. Journal of the Idaho Academy of Sciences, December.

Rugh, Scott, 1998. Clams of Champions: the San Diego Formation, http://www.sdnhm.org/research/paleontology/sdform.html

Segal, Ronald, 1965. New Fossil Fruit (Compositae) from the Pliocene of Western Kansas. American Midland Naturalist 73(2): 430-432.

Ch. 17: Pleistocene

Anonymous, 1999. Digging Deeply Into Paleontology, South Dakota School of Mines Quarterly, Summer issue.

—, 2004a. Pleistocene, http://en.wikipedia.org/wiki/Pleistocene

—, 2004b. Pleistocene Megafauna, http://www.bagheera.com/inthewild/ext_woollym.htm

—, 2007. Freshwater Mollusks from the Sehoo Formation, Nevada, http://members.aol.com/Waucoba7/ammonoids/ammonoids.html

Baskin, Jon A., 2004. The Pleistocene Fauna of South Texas, http://users.tamuk.edu/kfjab02/SOTXFAUN.htm

Clos, Lynne M., 1999. Where Have All the Mammals Gone? Fossil News 5(7): 3-8.

Columbia Encyclopedia, 2004. Pleistocene Epoch, http://www.bartleby.com/65/pl/Pleistoc.html

Cowen, Richard, 2005. History of Life, 4th edition. Blackwell Publishing, Oxford, UK.

Elias, Scott, 2004. Ice Age Explanation, http://culter.colorado.edu/~saelias/glacier.html

Guthrie, R. Dale, 2005. The Nature of Paleolithic Art. University of Chicago Press, Chicago, IL.

Illinois State Museum, 2004a. Why Were There Ice Ages? http://museum.state.il.us/exhibits/ice_ages/why_4_cool_periods.html

—, 2004b. The Late Pleistocene Extinctions, http://museum.state.il.us/exhibits/larson/LP_extinctions.html

—, 2004c. Human Hunting, http://museum.state.il.us/exhibits/larson/overkill.html

—, 2004d. Environmental Causes, http://museum.state.il.us/exhibits/larson/env_change_extinction.html

Johnson, Kirk, and Richard Stucky, 1995. Prehistoric Journey: a History of Life on Earth. Roberts Rinehart Publishers, Boulder, CO.

Kazlev, M. Alan, 2002. The Quaternary Period, http://www.palaeos.com/Cenozoic/Quaternary.htm

Kentucky State Parks, 2004. Big Bone Lick State Park, http://www.state.ky.us/agencies/parks/bigbone.htm

Kurtén, Björn, and Elaine Anderson, 1980. Pleistocene Mammals of North America. Columbia University Press, New York.

Lange, Ian M., 2002. Ice Age Mammals of North America. Mountain Press, Missoula, MT.

Mammoth Site, 2004. The Mammoth Site website, http://www.mammothsite.com

Natural History Museum of Los Angeles County, 2004. Return to the Ice Age: the La Brea Exploration Guide, http://www.tarpits.org

Ogg, James, 2003. Overview of Global Boundary Stratotype Sections and Points (GSSP's). http://www.stratigraphy.org/gssp.htm

Peterson, Morris S., J. Keith Rigby, and Lehi F. Hintze, 1973. Historical Geology of North America. Wm. C. Brown Co., Dubuque, IA.

Pielou, E. C., 1991. After the Ice Age. University of Chicago Press, Chicago, IL.

Potts, Richard and Anna K. Behrensmeyer (rapporteurs), 1992. Late Cenozoic Terrestrial Ecosystems. in Behrensmeyer, Anna K. et al. (eds), Terrestrial Ecosystems Through Time. University of Chicago Press, Chicago, IL.

Purdue University, 2004a. Pleistocene Epoch, http://agen521.www.ecn.purdue.edu/AGEN521/epadir/wetlands/pleistocene.html

—, 2004b. Taiga, http://agen521.www.ecn.purdue.edu/AGEN521/epadir/wetlands/taiga.html

—, 2004c. Tundra, http://agen521.www.ecn.purdue.edu/AGEN521/epadir/wetlands/tundra.html

—2004d. Kettleholes, http://agen521.www.ecn.purdue.edu/AGEN521/epadir/wetlands/kettleholes.html

Roy, K., 2001. Pleistocene Extinctions. in Paleobiology II, Briggs, D. and Crowther, P. (eds), Blackwell Science, Oxford, UK.

Sanders, Robert, 2003. Colorado Cave Yields Million-Year-Old Record of Evolution and Climate Change, UC Berkeley press release 10/21/03.

Scotese, Christopher R., 2004. The Earth Has Been in an Ice House Climate for the Last 30 Million Years, http://www.scotese.com/lastice.htm

Sinibaldi, Robert, 1998. Fossil Diving in Florida's Waters, or Any Other Waters Containing Prehistoric Treasures. Published by the author, 6458 29th Ave N, St. Petersburg, FL 33710.

Speer, Brian R., 1998. Localities of the Pleistocene: the La Brea Tar Pits, http://www.ucmp.berkeley.edu/quaternary/labrea.html

Teller, James T., 2005. Lake Agassiz Overflow During the Younger Dryas, http://gsa.confex.com/gsa/2005AM/finalprogram/abstract_92422.htm

University of Kentucky, 2004. Did You Know That Benjamin Franklin and Thomas Jefferson Studied the Fossil Bones from Big Bone Lick in Northern Kentucky?, http://www.uky.edu/KGS/education/bigbonelick.html

Ward, Peter D., 1997. The Call of Distant Mammoths. Copernicus Books, New York, NY.

Ch. 18: Holocene

Anonymous, 2007a. History of the Stratigraphical Nomenclature of the Glacial Period, http://www.quaternary.stratigraphy.org.uk/about/history.html

—, 2007b. Kyoto Protocol, http://en.wikipedia.org/wiki/Kyoto_Protocol

—, 2007c. Carbon dioxide, http://en.wikipedia.org/wiki/Carbon_dioxide

—, 2007d. Methane clathrate, http://en.wikipedia.org/wiki/Methane_clathrate

—, 2007e. Holocene, http://www.answers.com/topic/holocene?cat=technology

—, 2007f. Passenger Pigeon Memorial Hut, http://www.roadsideamerica.com/attract/OHCINmartha.html

Fagan, Brian, 2000. The Little Ice Age: How Climate Made History, 1300-1850. Basic Books, New York, NY.

Knutson, Tom, 2004. Climate Impact of Quadrupling CO_2, http://www.gfdl.noaa.gov/~tk/climate_dynamics/climate_impact_webpage.html

Lovgren, Stefan, 2004a. Greenland Melt May Swamp LA, Other Cities, Study Says. National Geographic News 4/8/04, http://news.nationalgeographic.com/news/2004/04/0408_040408_greenlandicemelt.html

—, 2004b. Warming to Cause Catastrophic Rise in Sea Level? National Geographic News 4/26/04, http://news.nationalgeographic.com/news/2004/04/0420_040420_earthday.html

Meehl, Gerald A. et al., 2005. How Much More Global Warming and Sea Level Rise? Science 307(5716): 1769-1772.

Ogg, James, 2003. Overview of Global Boundary Stratotype Sections and Points (GSSP's). http://www.stratigraphy.org/gssp.htm

Pielou, E. C., 1991. After the Ice Age. University of Chicago Press, Chicago, IL.

Roach, John, 2006. Greenland Glaciers Losing Ice Much Faster, Study Says. National Geographic News 2/16/06, http://news.nationalgeographic.com/news/2006/02/0216_060216_warming.html

Titus, James G. and Charlie Richman, 2001. Maps of Lands Vulnerable to Sea Level Rise. Climate Research, U.S. Environmental Protection Agency.

Index

USGS
science for a changing world

U.S. DEPARTMENT OF THE INTERIOR
U.S. GEOLOGICAL SURVEY

This map, which shows the surface form and age of bedrock across the North American continent, is a digital combination of two new maps: a shaded relief map and a geologic map. Geologic data were compiled by John C. Reed, Jr.[1] and John O. Wheeler[2], for the Decade of North American Geology Geologic Map of North America, sponsored by the Geological Society of America. These geologic data were compiled and generalized by David G. Howell[1]. The shaded relief map was created from 1-km resolution digital elevation data. This project, which follows the similar Tapestry of Time and Terrain for the conterminous United States (USGS Geologic Investigations Series I-2720), resulted from a collaborative effort among the Geological Survey of Canada, the United States Geological Survey, and the Consejo de Recursos Minerales de Mexico. We thank Richard Pike of the USGS for valuable feedback on drafts of this project: John Hutchinson of the EROS Data Center for the digital elevation model of North America and guidance on preparing the shaded relief map; and Jane Ciener of the USGS for editorial guidance.

Este mapa, el cual muestra la topografía y la edad de las rocas a lo largo de Norte América, representa la combinación digital de dos mapas: uno de relieve sombreado y un mapa geológico compilado recientemente. Los datos geológicos fueron compilados por John C. Reed, Jr.[1] y John O. Wheeler[2], para la Década de la Geología de Norte América y mapa geológico de Norte América, patrocinada por la Sociedad Geológica de América. La generalización de la geología y su adaptación fue realizada por David G. Howell[1]. Este proyecto es en seguimiento de la Cobertura de Tiempo y Terreno, el cual cubrió solamente el interior de los Estados Unidos. Este proyecto representa un esfuerzo de colaboración entre El Servicio Geológico de Canada, El Servicio Geológico de Los Estados Unidos, y el Consejo de Recursos Minerales de México. Queremos agradecer, en particular, a Richard Pike del USGS, quien proporcionó un apoyo valioso sobre los borradores de este proyecto: John Hutchinson de Eros Data Center, quien proporcionó el modelo digital del terreno de Norte América y oriento para la creación del mapa de relieve sombreado; así como a Jane Ciener del USGS, quien proporcionó la guía editorial.

Cette carte, qui illustre la topographie et les âges du socle rocheux de l'Amérique du Nord, représente une combinaison numérique de deux cartes, l'une de la topographie à relief par ombres portées, et l'autre de la géologie de compilation récente. Les données géologiques ont été compilées par John C. Reed, Jr.[1] et John O. Wheeler[2] pour la carte Geologic Map of North America, dans le cadre de la Decade of North American Geology, parrainée par la Geological Society of America. La généralisation et l'adaptation de la géologie sont une réalisation de David G. Howell[1]. Le projet fait suite au Tapestry of Time and Terrain, qui ne couvrait que les États contigus des États-Unis d'Amérique. Il s'agit d'une collaboration de la Commission géologique du Canada, de la United States Geological Survey (USGS) et du Consejo de Recursos Minerales du Mexique. Nous tenons à remercier en particulier Richard Pike (USGS), qui a fourni des conseils précieux sur les premières ébauches du projet; John Hutchinson (EROS Data Center), qui a fourni le modèle altimétrique numérique de l'Amérique du Nord ainsi que des conseils sur la création de la carte topographique à relief par ombres portées; et Jane Ciener (USGS), qui a fourni des conseils en matière de rédaction.

Hawaiian Islands

Geologic Time Scale
Escala de tiempo geológico
Échelle des temps géologiques

Millions of Years Ago (non-linear)
Millones de años (escala no lineal)
Millions d'années dans le passé (échelle non linéaire)

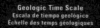

Quaternary - Cuaternario - Quaternaire

Neogene - Neógeno - Néogène

Paleogene - Paleógeno - Paléogène

Cretaceous - Cretácico - Crétacé

Jurassic - Jurásico - Jurassique

Triassic - Triásico - Trias

Permian - Pérmico - Permien

Pennsylvanian - Pensilvánico - Pennsylvanien

Mississippian - Misisipico - Mississippien

Devonian - Devonico - Dévonien

Silurian - Silúrico - Silurien

Ordovician - Ordovicico - Ordovicien

Cambrian - Cámbrico - Cambrien

Late Proterozoic - Proterozoico tardío - Protérozoïque tardif

Middle Proterozoic - Proterozoico medio - Protérozoïque moyen

Early Proterozoic - Proterozoico temprano - Protérozoïque précoce

Late Archean - Arqueano tardío - Archéen tardif

Middle Archean - Arqueano medio - Archéen moyen

Early Archean - Arqueano temprano - Archéen précoce

Glacial ice - Glaciar - Glacier

Age unknown - Edad desconocida - Âge inconnu

Geological Survey of Canada
Commission géologique du Canada

Consejo de Recursos Minerales